Landmarks on the Iron Road

Railroads Past and Present

Edited by George M. Smerk

Landmarks on the Iron Road

Two Centuries of North American Railroad Engineering

WILLIAM D. MIDDLETON

Indiana University Press

Bloomington and Indianapolis

This book is a publication of

Indiana University Press
601 North Morton Street
Bloomington, Indiana 47404-3797 USA

iupress.indiana.edu

Telephone orders   800-842-6796
Fax orders   812-855-7931
Orders by e-mail   iuporder@indiana.edu

First paperback edition 2011
© 1999 by William D. Middleton
All rights reserved

No part of this book may be reproduced or utilized in any form or by any means, electronic or mechanical, including photocopying and recording, or by any information storage and retrieval system, without permission in writing from the publisher. The Association of American University Presses' Resolution on Permissions constitutes the only exception to this prohibition.

∞ The paper used in this publication meets the minimum requirements of the American National Standard for Information Sciences—Permanence of Paper for Printed Library Materials, ANSI Z39.48-1992.

MANUFACTURED IN THE UNITED STATES OF AMERICA

The Library of Congress has catalogued the hardcover edition as follows:

Middleton, William D., date
Landmarks on the iron road : two centuries of North American railroad engineering / William D. Middleton.
  p. cm. (Railroads past and present)
  Includes bibliographical references and index.
  ISBN 0-252-33559-0 (alk. paper)
  1. Railroads—United States—History. 2. Railroad engineering—History. I. Title. II. Series.
TF23.M52   1999
625.1' 00973—dc21   98-55475

ISBN 978-0-253-33559-3 (cloth)
ISBN 978-0-253-22360-9 (pbk.)

1 2 3 4 5 16 15 14 13 12 11

Contents

Preface vii

1
Across the Waters 1

2
Across Great Mountains 69

3
Railroads below Ground 103

4
Yards, Docks and Terminals 141

5
Lost Landmarks 171

Bibliography 187

Index 191

Preface

Railroad engineering began in England, but it was soon taken up by railroad builders on this side of the Atlantic and developed in uniquely American ways, suitable to the formidable tasks of building a railroad system across the huge North American continent. American civil engineers were soon unsurpassed in their ability to build railways over vast distances and across high mountain passes, to erect great bridges, or to bore tunnels of prodigious length. Theirs is a remarkable story of the inspired application of engineering to the building of a transportation system that civilized and settled America, and then supported an industrial revolution that made it a world power.

To a remarkable degree, too, the railroad nurtured the new profession of civil engineering in the nineteenth century. Schooled at West Point, the technical schools of Europe, or new technical institutes in America, and sometimes self-taught, these early civil engineers soon acquired the skills and experience needed for the location and construction of railroads that could be rapidly and economically built. Their innovation and daring developed the methods and machines that transformed tunneling from a hand drilling and black powder blasting art to a modern technology of shields and tunnel-boring machines. The needs of the railroads changed bridge design from a trial-and-error art to an analytical science, fathered modern structural engineering practice, and advanced the development of structural materials.

Aside from a proliferation of works on bridges, the extraordinary achievements of railroad engineers have received relatively little attention. For the industry's historians and enthusiasts, an interest in railways is most often manifested in a preoccupation with locomotives, cars and the trains themselves. The physical plant over which they operate—the roadbed, track, bridges and tunnels—is more often taken for granted, a sort of backdrop to the action. But even if this railroading infrastructure does not always engage our primary interest, it represents an element of the industry no less important than flanged-wheel-on-steel-rail technology itself.

Combining an abiding interest in railroad history with the professional interests of a civil engineer, I have long had a special fascination in these great works of civil engineering. In this volume I have set out to tell the stories of some of them and of the men who designed and built them.

In addition to the works of railroad location, bridges and tunnels that one would expect to find in a volume such as this, I have also included examples of the remarkable engineered works that railroads built to load, unload or transship great quantities of grain, ore, coal, and the like; to classify freight cars; or to load and unload containers. Designed and built by engineers without any aesthetic pretenses, these huge industrial structures and yards often achieved an extraordinary visual power in their sheer size, functionality, and mechanical ingenuity.

The choice of which works to include was not always an easy one. Some were based upon the historical significance of a project in the development of the science of railroad engineering, while others were based upon the magnitude or importance of a project at the time of its completion. Still others were chosen as works representative of a particular type of structure or facility. A few were based simply on personal preference or interest. Within each chapter they are included in chronological order of their completion date and are listed under the name of the owning company at that time.

In developing this story of engineering achievement I have been aided by a rich historical record, both in contemporary accounts of many of the individual projects at the time they were built and in more recently published works. In the bibliography I have included a brief listing of relatively recent works that are sources of additional information about the individual works, a number of more general sources that I considered particularly worthwhile, and a general guide that will direct the reader toward some of the many sources of technical information that were contemporary with each of the works included in this volume.

As always, I would not have been able to successfully complete a project of this kind without the able and willing assistance of a great many institutions, corporations, and individuals.

A number of libraries, archives and historical societies have been of particular assistance in locating a wide variety of reference materials and illustrations. Archivist John P. Hankey of the Baltimore & Ohio Museum at Baltimore, Maryland, helped with illustrations of historic B&O bridges. The Hagley Museum and Library at Wilmington, Delaware, provided valuable assistance concerning Pennsylvania Railroad and Reading topics from both photographic resources and archival material. The Hagley's Barbara Hall of the pictorial collections and archivist Christopher T. Baer were particularly helpful. Librarians Barbara Gill at the Historical Society of Berks County (Pennsylvania) and Marion Strode at the Chester County (Pennsylvania) Historical Society helped locate photographs and historical material concerning the Reading's pioneering Black Rock Tunnel. David Pierce of the Chester (Massachusetts) Foundation provided some valuable background material on the history of George Washington Whistler's pioneering construction of the Western Railroad through the Berkshires. Marion G. Sweet and the Reverend Garford F. Williams of the Nicholson (Pennsylvania) Heritage Association provided similar help with material concerning Tunkhannock Viaduct.

In Washington, D.C., the staff of the Prints and Photographs Division of the Library of Congress were unfailingly helpful in making available material from the Library's vast holdings. William W. Worthington, Jr., museum specialist in the Smithsonian Institution's division of engineering and industries, helped to locate valued reference material and illustrations in the division's extensive files on engineering projects.

Without fail the Linda Hall Library at Kansas City offered exemplary service in providing copies from the library's extensive holdings in technical journals. At the Northeast Minnesota Historical Center in Duluth, administrator and manuscripts curator Patricia Maus helped with research material and illustration of Missabe Road ore docks. The Rensselaer Polytechnic Institute Archives at Troy, New York, made available copies of drawings and other materials from their extensive Roebling collections. At Minneapolis, David Wickins of the St. Anthony Falls Interpretive Center provided useful material concerning the Great Northern's stone arch viaduct. Carolyn A. Davis, manuscripts librarian at Syracuse University's George Arents Research Library, helped locate several splendid glass plate negatives of the Lackawanna's Tunkhannock Viaduct. The staff at the Visual Material Archives at the State Historical Society of Wisconsin assisted in finding rare illustration of the Milwaukee Road's Prairie du Chien floating drawbridge. At Charlottesville, the staffs of the University of Virginia's Science and Engineering and Alderman Research libraries have cheerfully filled an almost endless variety of research requests.

In Canada the British Columbia Archives at Victoria, the Glenbow Archives at Calgary, the National Archives of Canada at Ottawa, the Vancouver Public Library, and the City of Vancouver Archives all assisted in my search for illustration of Canadian topics. At the McCord Museum in Montreal curator of photographs L. Stanley Triggs located a number of important illustrations from the Notman Photographic Archives. Special collections librarian Debera Earle at the Hamilton (Ontario) Public Library was particularly helpful in locating material about the life and engineering career of Joseph Hobson.

In the United Kingdom, Miss L. Lean at the Science Museum, London, located a photograph of the distinguished British civil engineer Robert Stephenson.

From the American Society of Civil Engineers book acquisition editor Mary Grace Luke-Stefanchik and public relations assistant Tony Simione provided details of the Society's National Historic Civil Engineering Landmark Program, as well as photographs of early Society presidents.

A number of corporations, too, have helped in a variety of ways. At the American Bridge Company senior vice president Michael D. Flowers and assistant to the president R. A. Talbot provided a number of valuable drawings, photographs and other material about the Arthur Kill and other bridges built by the company. Libby Hunter at Atlas Copco Robbins, Inc., furnished material about the company's modern tunnel-boring machines. The Bethlehem Steel Company's assistant editor William J. Gignac kindly furnished similar material about that company's recent bridge projects.

At Burlington Northern Santa Fe, art and archives manager David C. Letourneau located a number of useful printed materials and photographs of the historic engineering projects of BNSF predecessor companies, while administrative assistant J. M. Hagen at the predecessor Burlington Northern assisted with extensive material concerning the Latah Creek bridge and the Great Northern stone arch bridge. At Canadian Pacific Railway, imaging specialist Rick Robinson and corporate archivists Stephen Lyons and the late Omer Lavallée provided a number of both historic and recent photographs of CP construction and tunneling in Kicking Horse and Rogers passes. Similar assistance was provided by news editor Donald N. Macintyre at Canadian National.

Jim Scribbins, then communications resources manager for the former Milwaukee Road, provided photographs, drawings and data concerning the Prairie du Chien and other company bridges. Rudy Husband, Conrail's media communications director, assisted with photographs and information concerning the railroad's main line tunnel enlargement project and Enola Yard. Duluth, Missabe & Iron Range manager of docks David G. Van Brunt provided much information that helped me to understand the present-day shiploader operation at the railroad's Dock No. 6 at Duluth. Miguel Tirado Rasso, gerente de comunicacion social at Ferrocarriles Nacionales de Mexico, provided both extensive information and photographs of FNM's modernization of the historic Mexico City–Vera Cruz line.

Janice M. Kitta, communications manager for HNTB Architects Engineers Planners, provided information and photographs of the award-winning Latah Creek bridge designed for BN by the firm. At Norfolk Southern, corporate communications director Richard W. Harris, agent terminal control Ervin Mullins, and assistant superintendent George Wilkie have all been exceedingly helpful in providing information and illustration concerning Spencer Yard and Coal Pier No. 6 at Norfolk, as well as a first-hand look at coal pier operations. The Pittsburgh & Lake Erie's operations vice president F. J. Habic and Paul Brown of the railroad's Pittsburgh & Conneaut Dock Company provided both data and photographs and kindly arranged an informative tour of dock operations at Conneaut.

Susan Gartrell, community affairs representative for Tri-Met's West Side Light Rail Project at Portland, Oregon, helped with photographs of that project's state-of-the-art tunnel boring machine. At Union Pacific public relations staff members Barry B. Combs, Michael J. Furtney, Mark Davis, and Edward J. Trandahl, as well as news editor William M. Robertson of predecessor Southern Pacific, have all been helpful in diverse ways. Finally, I am indebted to project engineer Ann Barker and public affairs director Linda Morris of the Vancouver Port Corporation, and to Terminal Systems, Inc. marketing director Morley Strachan for information, photographs and a behind-the-scenes tour of the latest in intermodal terminal technology at Vancouver's Deltaport.

I am indebted to a great many individuals who helped in many diverse ways. Photographer Michael Bailey copied a number of rare illustrations with great skill, and Stan Kistler produced prints of exceptional quality from a wide variety of negatives. Lawrence R. Duffee assisted with data concerning Richmond, Fredericksburg & Potomac bridges. Preston P. Dunavant, Jr., the retired assistant chief engineer, construction for the Norfolk & Western, shared with me a wealth of information from his experience in planning and constructing the railroad's Coal Pier No. 6 at Norfolk. Harold B. Hall and John D. Mitchell, Jr. provided much help to the story of the Burlington's Metropolis bridge from their encyclopedic knowledge of that historic span. Joe Piersen and Paul Swanson of the Chicago & North Western Historical Society were of similar help to me in completing the account of the North Western's prodigious classification yard at Proviso. William Sepe, chairman of the Poughkeepsie-Highland Railroad Bridge Company, has kept me posted on the status of that organization's innovative plans for the historic Poughkeepsie bridge. David N. Skillings, Jr., publisher of *Skillings Mining Review*, provided some excellent illustration of the Missabe Road's modified Dock No. 6 at Duluth. Edwin C. Storey helped with extensive material concerning the Erie's long-vanished timber bridge on the Genesee River at Portage, New York.

For a wide range of advice, encouragement and assistance with both historical materials and illustration I am in debt, too, to Golden West Books publisher Donald Duke, Prof. H. Roger Grant of Clemson University, Herbert H. Harwood, Jr., Prof. Don L. Hofsommer of St. Cloud State University, F. H. "Joe" Howard, Louis M. Newton, Jim Shaughnessy, and J. W. Swanberg.

At Kalmbach Publishing Company Kevin P. Keefe, editor of *Trains* magazine, has kindly assisted by making available some key illustrations from the *Trains* collection held in the David P. Morgan Memorial Library, while librarian Nancy L. Bartol was of much help in my research efforts. George H. Drury, now retired as a Kalmbach Books acquisitions editor, helped formulate the original concept for the book, and has been unfailing in his encouragement.

Finally, I acknowledge with special appreciation the contributions of my wife, Dorothy, whose advice and critiques have done much to enhance the clarity of the manuscript, and of my eldest son, William D. Middleton, III, who expertly translated from the Spanish important materials concerning the Mexico City–Vera Cruz railroad. As always, I reserve to myself responsibility for any errors or omissions they may have overlooked.

Landmarks on the Iron Road

Erecting the structure before it as it went, a track-mounted crane completed a viaduct across Kelley Creek in Idaho on the Milwaukee Road's Pacific Extension in 1908.—H. English Photo, Library of Congress (Neg. LC-USZ62-74929).

1
Across the Waters

Central to the achievements of America's nineteenth-century era of railroad expansion were the great bridges of stone, wood, iron, and steel that carried the rails across the wide rivers and deep valleys of the North American continent. The art of bridge building had been practiced for centuries before the advent of the railroad, but well into the nineteenth century it remained a tedious, empirical process that was ill suited to the needs of the railroads for quickly built, economical structures that could safely carry the heavy loads of locomotives and cars. Indeed, it took the needs of the railroads to advance bridge building to a scientific process and to lay much of the foundation for the modern profession of civil engineering.

One of the oldest bridge forms was the masonry arch, which dated to prehistoric times; and the builders of the Baltimore & Ohio and other pioneer railroads turned to this for their first major bridges. These large arches were built in place. Temporary timber "centering" was erected to support each arch until it was complete. Stone was quarried as close to the site as it could be found and then cut and mortared into place. The great strength and durability of these early arches were evidenced by their long service lives. Many still serve today. But the skilled craftsmen needed to quarry, cut, and lay the stone were always in short supply; and the building of masonry structures was a slow and costly process. It soon became evident that other structural forms were needed to provide bridges that could be built quickly and cheaply to keep pace with the growth of the railroad network.

The railroad builders turned first to timber structures. Wood was plentiful and cheap, and it was easily shaped and worked. Arched timber truss road bridges with spans of as much as 360 feet had been built as early as 1804. In 1830 the B&O engaged Lewis Wernwag, one of the most successful builders of these pioneer timber bridges, to build the first wood-framed railroad bridge in America. This was a 350-foot-long bridge of three arched deck trusses over the Monocacy River in Maryland. Several years later Wernwag built the B&O's first crossing of the Potomac at Harpers Ferry, a seven-span timber structure that was 830 feet long.

The arched designs used for most of these early timber bridges were not well suited to most railroad uses, since they required heavy abutments capable of resisting the horizontal thrust of the arches. Moreover, the arch spans made it more difficult to provide the level floor required for tracks. Simple truss bridges, on the other hand, readily provided a level deck and exerted only vertical loads at their supports.

One early design favored by railroad builders was a wood lattice truss patented by Ithiel Town in 1820. This was built with horizontal top and bottom chords of timber members with a latticework web of planks between them. Most were built as covered bridges, contributing to a remarkable longevity. The design was popular in New England, where more than a hundred survived into the twentieth century on Boston & Maine branches.

A much more successful early timber truss design was developed in 1838 by William Howe of Spencer, Massachusetts. The Howe truss was not, strictly speaking, an all-timber design, since it combined heavy top and bottom chords and diagonal members of wood with vertical tension members of wrought iron. Later simplified, the Howe design soon became the most widely used wooden truss for railroad bridges.

While wood remained popular for railroad bridges through much of the nineteenth century, bridge designers were beginning to use metal even before the beginning of the railroad era. Typically, these early metal bridges were built of cast or wrought iron. The brittle cast iron was usually used only for members under compressive loads, while the tougher wrought iron was used for tension members.

The first metal railroad bridge in North America was built by Richard B. Osborne, chief engineer for the Philadelphia & Reading, at Manayunk, Pennsylvania, in 1845. This was a 34-foot Howe truss span, in which cast iron members replaced the usual wooden diagonals and wrought iron sections were used for the vertical members and the top and bottom chords. One of the trusses from this historic bridge survives today in the Smithsonian Institution's National Museum of American History at Washington.

The early railroad builders had plenty of other truss designs to choose from for iron bridges. In 1846 Squire Whipple, one of the foremost early bridge engineers, developed the Whipple truss, which was made from cast and wrought iron. At about the same time Boston engineer Thomas W. Pratt and his father, Caleb, designed the Pratt truss, which reversed the web design of the Howe truss to place the diagonal members in tension and the vertical members in compression. One of the most popular truss forms was the Warren truss, first built in 1846 by British engineers James Warren and Willoughby Monzani. This was designed with diagonals that were alternately sloped in opposite directions, one in tension and one in compression. In the original design there were no vertical posts, but these were added in later versions.

Still other popular truss designs were developed by two of the early B&O engineers. In 1850 Wendel Bollman developed a type of suspension truss which employed cast iron top and bottom chords, wrought iron web diagonals, and a system of wrought iron suspension rods radiating from the top of the end posts. Another type of suspension truss was developed a year later by Albert Fink. Usually built as a deck truss, the Fink design was built with a top chord and posts of either wood or cast iron and a series of wrought iron diagonals in tension radiated from the ends to the bottom of each post. The trusses were usually built without a bottom chord, and additional diagonals took the tension normally carried by the chord. The Fink truss was exten-

A notable example of Ithiel Town's patented wood lattice truss design was the Richmond & Petersburg's crossing of the James River at Richmond, Virginia. Completed in 1838, the 2900-foot bridge had 19 spans ranging from 130 to 153 feet in length.—Author's collection.

sively used on the B&O's western extension from Cumberland, Maryland, and was later used for some of the most notable long-span bridges of the time.

Through the middle of the century the design of these bridges remained largely empirical in nature, and bridge builders had little to guide them as they tried to cope with steadily increasing locomotive weights and operating speeds. "The science of bridge proportioning was yet undeveloped," wrote the noted bridge engineer Theodore Cooper in his classic treatise *American Railroad Bridges*. "The best that the engineer could do, was to make the bridges stronger than heretofore, solely on the facts brought out by past experience."

But paralleling the increasing use of metal for bridge construction was the development of scientific methods for the analysis of stresses and the proportioning of members in a bridge structure. The first American engineers to develop a thorough understanding of stress analysis in trusses were Squire Whipple and Herman Haupt, who independently developed and published their theories in 1847 and 1851, respectively.

The work of Whipple, Haupt, and others marked the beginning of an era of scientific bridge analysis and design and a period of remarkable progress in bridge building. In 1859, for example, Albert Fink completed a remarkable bridge over the Green River south of Louisville on the Louisville & Nashville main line. With two truss spans of 180 feet and three of 208 feet, it ranked as the longest iron bridge in America.

Fink's record-breaking span length was soon surpassed by crossings of the Ohio River completed over the next two decades. By 1877 Jacob H. Linville had pushed

Sturdy and easily fabricated, the Howe truss was easily the most popular truss form in the era of wooden bridges. Typical of Howe truss spans was this crossing of the Kootenay River at Taghum, west of Nelson, British Columbia, on the Columbia & Kootenay, a Canadian Pacific subsidiary. —Glenbow Archives (File No. NA-118-42)

This handsome engraving by A. C. Warren depicted Albert Fink's greatest bridge, a crossing of the Ohio River at Louisville for the Pennsylvania Railroad completed in 1870. The span incorporated 25 Fink truss deck spans averaging 180 feet in length and through truss channel spans of 370 and 400 feet. These employed a new truss designed by Fink with subdivided panels which, in slightly different form, became known as the Baltimore, Petit, or Pennsylvania truss that was widely used for long-span bridges.—Author's collection.

Across the Waters   3

The Northern Pacific's crossing of the Missouri River at Bismarck, Dakota Territory, employed three Whipple truss spans that were 50 feet deep and 400 feet long. Completed in 1882, the bridge was designed by George S. Morison and Charles C. Schneider, two leading bridge designers of the late nineteenth century. Until the great bridge was completed, NP trains had to cross the Missouri by transfer boats or, in the winter, on the ice. This drawing from Scribner's *The American Railway* (1892) shows a test of the center span with a string of eight locomotives.—Author's collection.

truss spans to a record 517 feet with his design for the main channel span of a crossing of the Ohio at Cincinnati for the Cincinnati & Southern.

But these spectacular achievements of the bridge builders in this era of railroad expansion were not without their price, for failure of their wooden and iron bridges—frequently with calamitous loss of life--was altogether too common. In the decade between 1870 and 1880, for example, railroad and highway bridges collapsed at a rate of about forty a year. One particularly disastrous failure claimed wide public attention. This was the collapse of the Lake Shore & Michigan Southern's bridge over the deep gorge of Ashtabula Creek at Ashtabula, Ohio, on the wintry night of December 29, 1876. One of two locomotives and all eleven cars of the railroad's *Pacific Express* fell with the bridge, and a total of 80 people died, either in the crash or the holocaust that followed when heating stoves set fire to the splintered wreckage of the wooden cars.

Investigation of this and other bridge failures revealed many shortcomings in railroad bridge building practice of the time. Frequently, designers had an imperfect understanding of the forces acting on a bridge structure or of the potential dangers against which they had to safeguard their structures. Steadily increasing loads often subjected bridges to stresses for which they had never been designed. Wood was subject to decay and fire, while cast iron was often of uneven quality; and it was brittle and extremely weak in tension. All too commonly bridges were hastily built by the cheapest means that the prudence of their builders would allow. Quite generally, too, bridges were designed and built for the railroads by independent bridge companies on a competitive bid basis.

By this time, however, wrought iron, which was much tougher and more reliable than cast iron, was rapidly coming into common use. Jacob Linville had pioneered the development of testing machines which permitted "test to failure" of full-sized bridge members, enabling engineers to accurately determine the properties of materials and the performance of structural members.

A natural outgrowth of this new capability was the development of detailed specifications governing the required performance of materials and the standards of workmanship and construction for bridges. Thomas C.

Bridge failures were altogether too common in the early years of railroad expansion. One of the worst was the collapse of a Lake Shore & Michigan Southern bridge at Ashtabula, Ohio, in 1876 that plunged the *Pacific Express* into the gorge of Ashtabula Creek with great loss of life. A woodcut artist depicted the aftermath for *Frank Leslie's Illustrated Newspaper* of January 20, 1877.—Author's collection.

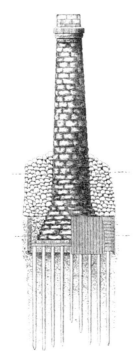

Where a solid foundation was unavailable at the bottom of a stream, bridge piers were usually carried on piling, which was either driven down to solid material or supported by friction with the soft material into which it was driven. This drawing from Scribner's *The American Railway* (1892) shows a typical pier founded on timber friction piling.—Author's collection.

Clarke and Samuel Reeves, partners in a bridge building firm, developed one of the first bridge specifications in 1871, and the practice was in general use by the end of the decade. Theodore Cooper, a young engineer who was to become one of the leading bridge engineers of the time, developed a specification for the Erie in 1878 that was the most comprehensive yet, and it became a model widely used by other roads.

At this time, too, engineers began to develop systems of concentrated loads approximating locomotive and train wheel loadings, which provided a far more accurate basis for the design of individual bridge members. Louis F. G. Bouscaren, chief engineer of the Cincinnati Southern, developed the first one in 1873, while Theodore Cooper later devised an improved system that was almost universally adopted by American railroads.

The converter process for making steel, developed in 1856 by English engineer and inventor Henry Bessemer, provided the basis for further advances in railroad bridge building. Steel possessed much greater strength, toughness, and wear resistance than either cast or wrought iron. Within a few decades the Bessemer process had made steel both plentiful and cheap, and it soon was widely used for bridge construction. James Buchanan Eads was able to use steel for the arch sections of his great Mississippi River bridge at St. Louis, completed during 1867–74; and the first all-steel bridge was erected over the Missouri River at Glasgow, Missouri, in 1879 by the Chicago & Alton. At first steel was typically used in combination with wrought iron, but by the 1890s bridges were usually being built entirely of steel.

Paralleling the rapid gains of the 1860s and 1870s in the materials and techniques of bridge design and construction were some notable advances in the field of foundation construction. In the early years of bridge building, cofferdams were normally used whenever it was necessary to construct foundations below water level. Typically, the cofferdam was an enclosure made of sheet piling driven into the stream bed or a box-like structure that was floated into place, sunk to the bottom of the stream, and then pumped out to enable construction of a pier "in the dry." Both the heavy loadings imposed on the cofferdam structure by deep water and the difficulty in pumping water out as the water depth—and pressure—increased limited the practical maximum depth for this method.

An alternative to the cofferdam was the pneumatic caisson. With this method excavation was carried out from inside what amounted to an upside-down box from which water was excluded by air pressure. Although used in Europe as early as 1850, its first major use in North America was by Eads during 1869-71 for the construction of foundations for the St. Louis bridge, where the deepest caisson was sunk to a depth of 136 feet below high water level. Eads encountered the new problem of "caisson disease," which was attributed to too rapid decompression as workers returned to the surface from the pressurized caissons. Once this was solved, the pneumatic caisson method helped to make possible many of the great bridges of the late nineteenth and early twentieth century.

With the availability of steel, bridge builders successfully pushed truss bridges to still greater span lengths. By 1895 William H. Burr had designed a Big Four bridge over the Ohio at Louisville that incorporated three truss spans each 547 feet long. Over the next two decades this was eclipsed by spans of 668 feet for the Municipal Bridge over the Mississippi at St. Louis, completed in 1912, and the 723-foot main span of the Paducah & Illinois crossing of the Ohio at Metropolis, Illinois, completed in 1917. But this was just about at the practical limit for the length of a simple, single-span truss; and the main span of the Metropolis bridge has never been exceeded.

Railroad truss bridges were usually built as simple spans, which spanned between two supports with an independent structure that was easily designed and fabricated. Because a simple truss bridge was incapable of taking any load until the structure was complete, these were usually erected on temporary falsework. Often,

Another drawing from *The American Railway* shows sandhogs at work inside a pneumatic caisson. Excavation work under pressure inside what amounted to an upside down wooden box 50 feet below the river's surface was difficult and dangerous.—Author's collection.

Wherever possible, bridge builders preferred to erect simple truss spans from temporary falsework. This 1915 view of the Phoenix Bridge Company's erection of a 920-foot truss for a new crossing of Occoquan Creek in Alexandria County, Virginia, for the Richmond, Fredericksburg & Potomac is typical of the practice. The span was designed by noted bridge engineer Gustav Lindenthal.—Hagley Museum and Library.

For spans over open water, where navigational requirements sometimes precluded temporary falsework, truss bridges were often built elsewhere and floated into position on barges. That was how the American Bridge Company erected this replacement truss span for the Pennsylvania's Delair Bridge across the Delaware River at Philadelphia in 1960.—Cliff Loane, American Bridge Company.

when a bridge spanned open water, trusses were assembled elsewhere and floated into position on a barge. But as engineers attempted to cross wider and deeper streams, the erection of these simple truss spans became increasingly difficult.

American bridge builders had begun at an early date to seek other structural forms that could provide more economical or practical solutions to the need for long clear spans. One of the earliest of these to be tried was the suspension bridge, a form that had been in use for centuries. Typically, these used cables or chains, supported by towers and anchored at each end of the span, to carry a suspended roadway; and they could be erected without any temporary falsework by first building the towers and anchorages, spinning the cables, and then hanging the roadway from the cables. The first railroad suspension bridge was erected across the Niagara River by John A. Roebling during 1851–55. With a clear span of 821 feet, it was the longest railroad bridge anywhere. But the lack of rigidity that was characteristic of suspension bridges proved a serious shortcoming for railroad use. On the Niagara bridge, for example, it was necessary to severely limit train speeds for this reason; and no other railroad suspension bridges were ever built in North America.

Another early bridge form that proved much more successful for long-span railroad needs was the metal arch. Cast iron arches had been built for road bridges in England as early as 1781, but the first metal arch railroad bridge in the U.S. was the St. Louis Bridge, completed by James Eads in 1874. This was followed by a number of long-span steel arch bridges, culminating in Gustav Lindenthal's record Hell Gate Bridge over the East River at New York, completed in 1917.

The use of arch bridges was generally limited to locations at which the site could provide foundations that could withstand the great horizontal thrust of the arches. A much more versatile structural form for extremely long railroad spans was the cantilever truss bridge, a bridge form first used in 1869 by German engineer Heinrich Gerber for a crossing of the Main River. In its basic form, the cantilever truss was typically made up of anchor arms at each end which were continuous with cantilever arms projecting beyond supporting piers. These two cantilever arms were then joined by a "suspended" or "floating" simple truss span at the center to form the main span of the bridge. With this arrangement, the anchor arms of the bridge counterbalanced the load of the cantilever arms and the suspended span, substantially reducing the maximum bending load at the center of the span and allowing significant material economies.

While the anchor arms at each end of a cantilever bridge were typically erected on falsework, much like a conventional truss bridge, the cantilever arms could be erected without it by building them outward from the supporting piers. The central suspended span was often erected by continuing to build outward from the end of each cantilever arm as a cantilever and then converting the center structure to a simple truss after the two halves were joined. Sometimes, too, the suspended span was assembled elsewhere, floated into position on barges, and lifted into place.

The first North American cantilever bridge was a Cincinnati Southern crossing of the Kentucky River completed in 1877 (see page 28). Still longer cantilever spans followed, and by 1892 engineers George Shattuck Morison and Alfred Noble had completed a crossing of the Mississippi River at Memphis, Tennessee, with a cantilevered span of 791 feet over the main channel of the river. But by 1917 this and all other cantilever spans were outdone by the completion of the great Quebec Bridge (see page 51), which remains today the longest cantilever bridge ever built.

Widely regarded as one of the most elegant examples of the steel arch form was this crossing of the St. Croix River east of St. Paul, Minnesota, on the Wisconsin Central. Designed by Claude A. P. Turner, the bridge incorporated five three-hinged arches, each with a rise of 124 feet and a span of 350 feet. The bridge was built during 1910-11 as part of a new cut-off between New Richmond, Wisconsin, and Withrow Junction, Minnesota, on the railroad's Chicago-Minneapolis line.—William D. Middleton.

One of the most important early cantilever bridges was this steel and wrought iron crossing of the Niagara River completed in 1883 for the Michigan Central by Charles C. Schneider. The anchor arms at either end of the bridge were 200 feet long, while cantilever arms of 175 feet and a 120-foot suspended span gave the bridge a clear span of 470 feet. The base of rail was all of 239 feet above water level. The drawing is from the December 1, 1883, issue of *Scientific American*.—Library of Congress.

Viaducts were easily erected from already completed sections of the span. This view shows Phoenix Bridge Company crews erecting the last 54-ton plate girder spans on the Central Railroad of New Jersey's Hometown Viaduct.—Hagley Museum and Library.

Still another bridge form that offered advantages over simple truss spans was the continuous truss, which employed a single truss structure that was continuous over two or more spans. This continuity, much like that of the cantilever, enabled one span to counterbalance the load of an adjacent span. It allowed significant economies in material over simple truss structures and made much longer clear spans feasible. Like the cantilever, too, continuous trusses were easily erected without falsework by the cantilevering method.

Continuous bridges had been built in England as early as 1850. The first in North America was a Canadian Pacific bridge over the St. Lawrence River near Montreal, designed by C. Shaler Smith and completed in 1887. However, the form was little used until Gustav Lindenthal completed an enormous bridge across the Ohio River at Sciotoville, Ohio, in 1917 (see page 57). This successful use of the continuous truss form led to its adoption for several other major North American bridges.

Still another metal bridge form, used as early as

The development of reinforced concrete arch construction early in the century produced some bridges of uncommon beauty. Designed by J. E. Greiner, this 761-foot arch span over the Rappahannock River was completed in 1927 as part of a project to elevate and double-track the Richmond, Fredericksburg & Potomac's line through Fredericksburg, Virginia. Span lengths of the structure's ten arches varied from 46 feet 8 inches to 103 feet 6 inches. Amtrak's southbound train 79, the *Carolinian*, crossed it in September 1993.—William D. Middleton.

1847 for a bridge on the Baltimore & Susquehanna, was the plate girder, which was widely employed for spans of moderate length, usually up to around 125 feet. Assembled from structural steel shapes and plates by riveting—and later welding—the plate girder is simply an "I" section of larger size than can be manufactured in a single rolled section.

Late in the nineteenth century there was a modest revival in the use of the masonry arch for railroad bridges, largely because of its permanence and its ability to permit train operation without speed restrictions. During this period both the Pennsylvania and the B&O built a large number of arch bridges, using both stone and Portland cement concrete. While these bridges used concrete largely as a substitute for stone masonry, the use of reinforced concrete for long span arches became still another popular bridge form early in the twentieth century. The Delaware, Lackawanna & Western was the leading practitioner of reinforced concrete arch construction, with a number of major structures built between 1908 and 1915 for extensive line relocations in New Jersey and Pennsylvania.

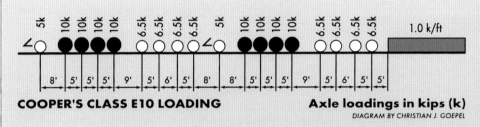

Cooper's E10 loading. Axle and uniform loadings of two 2-8-0 Consolidation locomotives and a following train are shown in units of kips (1000 pounds).—Drawing by Christian Goepel.

Mr. Cooper's Loading

Until the 1870s engineers typically designed bridge structures by assuming a uniform live load on the structure. This method provided design loadings that often varied widely from those actually produced by the concentrated axle loads of a heavy locomotive. Beginning with the work of Louis G. F. Bouscaren in 1873, railroads based their designs on a series of concentrated loads which more closely approximated the actual loads imposed by a locomotive. Initially there was wide diversity in the loading assumptions used by different railroads, but the system developed by bridge engineer Theodore Cooper was soon adopted as a standard loading system by most American railroads.

Cooper's loading assumes a freight train pulled by double-headed 2-8-0 Consolidation type locomotives of standard dimensions, followed by an indefinite number of freight cars. The axle loadings for the locomotives, and a uniform load representing the train, vary with the loading standard adopted for a particular line or structure. Cooper's Class E10 loading, for example, is based upon driving axle loadings of 10,000 pounds for the two Consolidation locomotives, and loadings of 5000 and 6500 pounds for the locomotive leading truck and tender axles respectively. The following train is represented by a uniform load of 1000 pounds per linear foot of track. These loadings are usually expressed in terms of kips, an engineer's "shorthand" unit representing 1000 pounds, rather than in pounds.

Different standards of loading are then based simply on multiples of this E10 loading. Axle and uniform loadings for a Cooper's E72 loading, for example, are 7.2 times the values for an E10 loading. This makes it possible to compute tables of bending moments (the bending effect, representing force times distance), shears, and floor beam reactions for an E10 loading that can readily be converted to the correct values for any class of loading by the use of a simple multiplier.

The loading standards used for railroad bridge design have steadily increased over the years as locomotive and train weights have grown. During the 1880s, when Cooper first introduced his loading system, bridges were usually designed for loadings no greater than Cooper's E20. By 1894 Cooper was recommending the use of his E40 loading as a standard, and within another 20 years bridges were commonly being designed for a Cooper's E55 or E60 loading standard. One of the heaviest loading standards ever adopted was used for the Burlington's bridge across the Ohio River at Metropolis, Illinois, which was designed for a live loading of two Cooper's E90 locomotives, followed by a uniform train loading of 7500 pounds per linear foot of track. For contemporary railroad bridge design the American Railway Engineering Association recommends the use of a Cooper's E72 loading for concrete bridges and an E80 loading for steel structures.

Nineteenth Century Bridge Engineers

Howe trusses of wood were widely used all over North America in the early period of railroad expansion. Toward the end of the wooden bridge era, Howe truss spans as long as 250 feet were built. These used a variation of the basic design similar to one patented by Howe in 1846 in which the trusses were reinforced by a pair of arch ribs, one on each side.

Squire Whipple (1804–1888) was among the most influential of the early railroad bridge engineers. Born at Hardwick, Massachusetts, he moved with his family to Otsego County, New York, as a youth, and studied at Union College in Schenectady. Following his graduation in 1830, Whipple began a career in engineering on the Baltimore & Ohio, where he was a surveyor during 1830–32.

Whipple's bridge building experience began in 1840, when he designed and patented a cast and wrought iron bow-string truss. Over the next decade, Whipple built a number of these bridges, which were said to be a majority of the iron bridges built before 1850. During 1852–53 he constructed a bridge spanning 146 feet for the Rensselaer & Saratoga, using a trapezoidal Whipple truss design that was widely adopted. His greatest contribution to American bridge engineering, however, was as the first American engineer to develop a thorough understanding of the principals of stress analysis in trusses. In 1847 his *A Work upon Bridge Building* set forth the first rational method of truss analysis. This influential work remained in print through a fourth edition published in 1883. He died at his home in Albany, New York.

William Howe (1803–1852) had an unlikely background for a bridge engineer, yet his design for a wooden truss was almost universally used by railroads in the United States and Europe. Born at Spencer, Massachusetts, Howe was working as a farmer when he was commissioned by the Western Railroad of Massachusetts to construct a bridge at Warren, Massachusetts, in 1838.

—Smithsonian Institution.

Howe designed and later patented a truss for this bridge that employed top and bottom chords and diagonal members, each extending across two panels, that were framed of heavy timber. The vertical tension members were made of wrought iron. In 1840 Howe joined with his brother-in-law, railroad builder Amasa Stone, Jr., to build the Western Railroad's bridge across the Connecticut River at Springfield, using a modified form of the Howe truss. The design was later further modified by Stone and others into a simplified form with a single diagonal across each panel, for which stresses could be calculated mathematically. Howe and Stone worked together for a time, and Stone later acquired Howe's patent rights and formed a company to build Howe truss bridges. Howe continued to build bridges and roof trusses until his death at Springfield in 1852.

—Smithsonian Institution.

Herman Haupt (1817–1905) was one of the most versatile and accomplished engineers of his time. A native of Philadelphia, he graduated from West Point in 1835 and soon left the Army to take up railroad location work. After teaching at Pennsylvania College in Gettysburg for several years, Haupt joined the Pennsylvania Railroad in 1847, later serving as its chief engineer during completion of the line over the Allegheny Mountains to Pittsburgh.

Haupt became interested in bridge construction early in his career and patented a design for a truss in 1839. He began a serious study of stress analysis in 1840, building models to test his theoretical designs. Soon after Squire Whipple had published his work on bridge design in 1847, Haupt completed his own book on the subject, *General Theory of Bridge Construction,* which was published in 1851 and was widely used by engineers for more than 30 years.

Haupt left the Pennsylvania in 1856 to take over construction of the Hoosac Tunnel (see page 111), an effort which finally defeated him. In 1862 he was called to Washington to take on the reconstruction of the Richmond, Fredericksburg & Potomac, and later, as a brigadier general, headed all Union military railroads. Following the war Haupt built more railroads and a petroleum pipeline, served as general manager of three railroads including the Northern Pacific, and was president of the Dakota & Great Southern. From 1886 until his death at Jersey City, New Jersey, he was involved in a variety of projects, chief among them an effort to develop compressed air power for street railway and rapid transit operation.

Albert Fink (1827–1897) was another of the leading nineteenth-century railroad engineers who emerged from the pioneer Baltimore & Ohio Railroad. The son of an architect, he was born at Lauterbach, Germany, and attended the Polytechnic School at Darmstadt, where he graduated with high honors in both architecture and engineering in 1848. Fink emigrated to the United States in 1849 and secured a position in the drafting office of Benjamin H. Latrobe, Jr., the B&O's chief engineer. Well-trained and capable, Fink soon became the principal assistant to Latrobe in charge of designing and erecting bridges and other structures for the expanding railroad.

—Louisville & Nashville Railroad.

While working under Latrobe, Fink designed the suspension truss that carried his name, and which was extensively used for long-span bridges by the B&O and other railroads. The longest of these were trusses spanning 208 feet for the Louisville & Nashville's crossing of the Green River south of Louisville. Later, Fink devised a new truss type for main channel spans of 370 and 400 feet for the Pennsylvania's crossing of the Ohio River at Louisville, completed during 1868–70. These used additional intermediate members to subdivide the truss panels. A simplified variation of the design, known as the Petit, Baltimore, or Pennsylvania truss, was later widely used by the Pennsylvania.

Fink left the B&O in 1857 to take up a post as assistant engineer for the L&N. After the Civil War, when he was in charge of reconstruction of sections of line destroyed by Confederate troops, Fink rose rapidly through L&N's management ranks, becoming general superintendent in 1865 and a vice president in 1870. After leaving the railroad in 1875, he headed two railroad associations concerned with through traffic, rates, and tariffs before retiring in 1888. He was president of the American Society of Civil Engineers during 1879–80.

Theodore Cooper (1839–1919) was one of the leaders in the late nineteenth century development of improved standards and practices for the design and construction of bridges. Born at Cooper's Plain, New York, he graduated in civil engineering from Rensselaer Polytechnic Institute at Troy in 1858. Cooper did survey work for the Troy & Greenfield and the Hoosac Tunnel before entering the Navy at the end of 1861. After Civil War service on board the *U.S.S. Chocura,* he was in charge of civil engineering work and taught at the Naval Academy.

—Rensselaer Polytechnic Institute Archives.

Cooper's experience in bridge design and construction began in 1872 when he left the Navy to work under Captain James Buchanan Eads in the construction of the St. Louis Bridge, where he was in charge of erection. His work on the landmark bridge was followed by a variety of posts in bridge and structures engineering before he established his own consulting practice at New York in 1879. Cooper developed an iron bridge specification for the Erie Railroad in 1878 that was the most comprehensive yet used and was adopted by many other railroads. This led to publication in 1884 of his widely used *General Specifications for Iron Railroad Bridges and Viaducts,* which went through seven editions by 1906. Cooper also developed the system of locomotive and train loading for bridge design that bears his name and which is still in use today. His *American Railroad Bridges,* published by the American Society of Civil Engineers in 1889, was a widely recognized history of the development of bridge engineering to that time.

In 1900 Cooper was appointed consulting engineer for the design and construction of a great cantilever bridge across the St. Lawrence River at Quebec. What was to have been the greatest achievement of his career ended instead in disaster when the bridge collapsed during construction with great loss of life in 1907. Censured by a Royal Commission for design errors, Cooper retired from engineering practice that same year.

The Baltimore & Ohio's Carrollton Viaduct (1829) at Baltimore, Maryland

Less than a year after the cornerstone was laid for the Baltimore & Ohio on July 4, 1828, America's first real railroad faced one of its first major engineering challenges at Gwynns Falls, a meandering creek in Baltimore only a short distance away from the site of the historic cornerstone itself. Colonel Stephen H. Long, one of the railroad's principal engineers, favored wooden bridges, and initially the B&O's board of directors had decided to bridge the stream with a wooden structure. But Caspar W. Wever, the road's superintendent of construction, persuaded them that a stone structure would be more durable and could be built economically from stone available at the site. The graceful granite arch that finally spanned Gwynns Falls would represent the first major engineered structure on an American railroad and the most enduring as well.

The Gwynns Falls structure was designed by Wever and built under his supervision, assisted by James Lloyd and John McCartney, who had previously built bridges for Wever in the construction of the National Road. Initially, Wever planned a single arch over Gwynns Falls of stone quarried on the site, but this quickly grew to a much larger and more costly structure owing largely to the requests of James Carroll, through whose property the line was being built. Carroll complained that the 50-foot arch planned by the B&O would not be big enough to handle the stream during floods. Wever then increased the span of the arch to 80 feet and added a 12-foot arch at one end to accommodate a road. Carroll thought that this wasn't big enough, and it was increased to a 20-foot arch. The prospective cost of the bridge went up once again when Wever decided to build it, not with the rough local stone originally contemplated, but with dressed granite hauled in by wagon from nearby Ellicotts Mills and Port Deposit, on the Susquehanna River. The bridge finally completed at the site was a massive segmental arch structure with spans of 80 feet and 20 feet, and was about 300 feet long, 26 feet 6 inches wide, and 58 feet high from the top of the foundation to the parapet.

A contract for construction of the bridge was awarded in December 1828, and John McCartney hauled in some lumber for the arch supports during that winter, but work did not begin on the site until the following May. Some 12,000 perches of stone went into the structure, and the builders erected temporary wood arch centering to support some 1500 tons of granite until the keystones were placed and the arch became capable of supporting itself. The completed structure was opened on December 21, when it was officially named the

This drawing of the Carrollton Viaduct appeared in an early issue of *Ballou's Pictorial Drawing-room Companion*.—B&O Railroad Museum.

Carrollton Viaduct. John Carroll of Carrollton rode out from the city in a horse-drawn train, laid the final stone, and then crossed the viaduct with a group of company and other officials.

Although it was acclaimed as a splendid structure, Carrollton Viaduct had proved a costly undertaking for the underfinanced B&O. In the end, owing to the many changes, it had cost the railroad at least four times the original cost estimate of about $15,000. The four-span stone arch Patterson Viaduct over the Patapsco River had proved a similarly costly venture, as would the great viaduct at Relay that would follow a few years later.

The B&O soon turned to less costly timber and iron structures as the railroad continued its push west toward the Ohio, but Carrollton Viaduct went on to confirm in full measure the durability claims of its builders. The bridge ceased to be part of the B&O main line in 1868 when the railroad completed a new route into downtown Baltimore, but it has remained part of an important freight line ever since. Today, CSX diesels rumble across the double-track span with freight trains many times heavier than could have been imagined when it was built. And it seems quite probable that this marvelous engineering work from the very dawn of the American railroad will still be carrying tonnage across the waters of Gwynns Creek when it comes time to celebrate its bicentennial. In 1971 the bridge was listed on the National Register of Historic Places, and in 1982 it was designated a National Historic Civil Engineering Landmark by the American Society of Civil Engineers.

GETTING THERE

Carrollton Viaduct is about 3 miles southwest of downtown Baltimore on Gwynns Falls, across the Carroll Park golf course from Washington Boulevard. A service road leads across the golf course to the viaduct.

The Baltimore & Ohio's Thomas Viaduct (1835) at Relay, Maryland

With its main line from Baltimore to the west well underway, the Baltimore & Ohio decided late in 1830 to build a branch to Washington. The most difficult part of the project by far was the construction of a viaduct across the valley of the Patapsco River at Relay, Maryland, where the branch diverged from the main line.

To design the structure the B&O selected Benjamin H. Latrobe, Jr., the youngest son of the great architect-engineer Benjamin H. Latrobe. The youthful (he was only 26 at the time) Latrobe proved as gifted as his father, and the viaduct at Relay would rank as one of the great works of early American railroad engineering.

The difficulty of Latrobe's task was compounded by the need to build the viaduct on a 4.5-degree curve that

At an age now approaching 170 years, Carrollton Viaduct continues to safely support heavy loads of modern railroading that could hardly have been imagined when the structure was completed in 1829. Two Chessie System 3000 h.p. Electro-Motive GP40-2 diesels headed a Baltimore & Ohio freight across the structure in September 1981.—William D. Middleton.

was required to join the branch to the main line. Latrobe designed a handsome stone masonry structure 704 feet long, including approaches, with eight elliptical arches, each spanning just over 58 feet. Because of the curvature, Latrobe laid out the faces of the piers on radial lines, giving them a wedge-shaped cross section, so that the arches did not have to be built on a skew. The roadway was 66 feet above water level, and Latrobe had the foresight to make it 26 feet wide, sufficient for a double-track line. Latrobe wanted to cap the structure with a stone parapet, but the railroad's directors insisted on a less expensive solution. Latrobe obliged them with a design for an elegant railing of cast and wrought iron.

Work began in August 1833, and for almost two years hundreds of laborers, carpenters, stonecutters, and stonemasons worked to erect the great viaduct. Several times the river flooded the work. Fifty feet of track were carried away by a landslide. A trestle collapsed, dropping three cars of stone into a millrace and killing a workman. Other workers were killed or injured in falls, or by falling construction materials. Digging the foundations for the piers through the hard clay of the river bottom proved tedious and time-consuming. Some 63,000 tons of granite for the structure were quarried at

The Baltimore & Ohio's Thomas Viaduct was one of the wonders of America in its time. Titled "Viaduct on Baltimore and Washington Railroad," this view from a drawing by artist W. H. Bartlett was engraved by H. Adlard and published at London in 1838. Visible to the left is the commemorative granite obelisk monument erected on completion of the structure in 1835 and still standing today.—Author's collection.

Thomas Viaduct was a favorite place for the B&O to photograph its newest equipment and trains. The new *Royal Blue* was eastbound on the viaduct with Class P-7 4-6-2 Pacific No. 5303 on April 16, 1937.—B&O Railroad Collection, Smithsonian Institution.

nearby Ellicotts Mills and hauled over spur lines and the railroad's main line to the site. There it was carried out over the valley on a temporary trestle and lowered by cranes into the cofferdams below.

In spite of the difficulties and the sheer magnitude of the work, the structure gradually took shape. By early November 1834 all eight arches were closed, and the viaduct was finally completed on July 4, 1835. As a final task, Latrobe designed a simple commemorative obelisk of granite, which still stands today at the north abutment. Named for the B&O's first president, Philip E. Thomas, the viaduct was placed in service the following August 25 with the formal opening of the Washington line. The largest bridge yet built in America and the first

14   **Landmarks on the Iron Road**

A principal present-day user of Thomas Viaduct is the Baltimore-Washington Camden Line of Maryland's busy MARC commuter rail service. Operating in push mode, MARC's westbound Baltimore-Washington train 257 crossed the venerable structure on August 25, 1989.—William D. Middleton.

ever constructed on a curve, Thomas Viaduct had cost the B&O $200,000, and it was to prove one of the most enduring works of American railroad engineering.

Despite its durability, the future of the viaduct was threatened as early as 1876, when the B&O began acquiring property for a cutoff between Halethorpe and Elkridge that would have eliminated the sharply curved viaduct and the curves on either side of it. Some work was started on the cutoff during the 1890s but was soon halted by the railroad's receivership. Deferred again and again, the cutoff project finally died for good with the sale of much of the land in 1972. Today the pioneering structure still carries heavy CSX and MARC commuter traffic over the Patapsco River Valley and has long since assumed the mantle of the oldest railroad bridge in main line service in North America. Thomas Viaduct was listed on the National Register of Historic Places in 1966.

GETTING THERE

Thomas Viaduct is at Relay, about 8 miles southwest of downtown Baltimore. It is most easily reached from U.S. 1 east of Elk Ridge. A road just east of the CSX underpass leads to the north under the west end of the viaduct, while a short distance further east an entrance road to Patapsco Valley State Park leads under the east end of the viaduct on the opposite bank of the river.

The Boston & Providence's Canton Viaduct (1835) at Canton, Massachusetts

Boston soon followed Baltimore's lead to develop its own railroads. One of the earliest was the 44-mile Boston & Providence, which began building south from Boston in 1834. Chief engineer Captain William Gibbs McNeil laid out an alignment for the new road that was as short and direct as possible, but it faced a formidable barrier near Canton, where the line would have to cross the deep valley of the east branch of the Neponset River.

McNeil designed a massive stone viaduct for the crossing that was 615 feet long and stood 70 feet above the surface of the water. Laid on a 1-degree curve, the structure crossed the river at a factory mill pond. Six semicircular arches, each spanning 8 feet 4 inches, provided a passage for the river, while a single 22-foot semicircular arch provided space for a roadway near the south end. The viaduct stood on a solid stone foundation. The hollow superstructure was made up of parallel stone walls separated by a 4-foot space and extending the full length of the viaduct, each wall about 5 feet thick at the base and battered slightly on the exterior face. The two walls were connected by occasional tie-stones, and at intervals of 27 feet 6 inches by transverse buttresses

—Smithsonian Institution.

Benjamin Henry Latrobe, Jr. (1806–1878), was born at Wilmington, Delaware. He attended Georgetown College in Washington and St. Mary's College in Baltimore before reading and practicing law with his brother John. In 1831 Latrobe joined the B&O as the principal assistant to chief engineer Jonathan Knight, and he was appointed the following year to design the Relay viaduct. He left the B&O following completion of the viaduct in 1835 to take up the post of chief engineer for the construction of the Baltimore & Port Deposit between Baltimore and Havre de Grace, Maryland. Latrobe returned to the B&O a year later to resume what was to be a long and distinguished engineering career with the railroad. After directing surveys and construction of the line west through Harpers Ferry and across the mountains to Cumberland, Maryland, he succeeded Knight as the B&O's chief engineer in 1842. He completed construction of the railroad to the Ohio River and built several other lines for the B&O. Still later Latrobe was a consulting engineer for the Hoosac Tunnel and the Portland & Ogdensburg. He retired in 1875 and died at Baltimore three years later.

A proud Boston & Providence featured this early drawing of the Canton Viaduct on an 1838 stock certificate.—Smithsonian Institution.

William Gibbs McNeil (1801–1853) was one of a cadre of U.S. Army engineers who helped to build the pioneer B&O and then moved on to build other early American railroads. Born at Wilmington, North Carolina, McNeil graduated from West Point in 1817. He was assigned to the Corps of Topographical Engineers for surveys of the Atlantic coast and the Gulf of Mexico and then served as aide-de-camp to Andrew Jackson during the Seminole War. During 1824–27 McNeil was assigned to survey work for several canals and then to the B&O. He joined Jonathan Knight and George Washington Whistler, his future brother-in-law, on a trip to study railway engineering practice in England before beginning work on the B&O. Over the next ten years he helped to build more than a half dozen other railroads, including the Boston & Providence and the Western Railroad of Massachusetts. McNeil left the Army in 1837 and continued a distinguished engineering career on a wide range of railroad, canal, and other projects in the United States, Canada, and the West Indies until his death in Brooklyn. He was the first American engineer honored by appointment to Great Britain's Institution of Civil Engineers.

that were 5 feet 6 inches thick. Each of these buttresses projected 4 feet beyond the wall faces on either side, forming pilasters which carried segmental arches spanning between them outside the faces of the wall to support the 22-foot-wide top of the viaduct, which carried a single track. A great earthen embankment more than 60 feet high completed the crossing of the valley at the north end of the viaduct.

Construction began on April 20, 1834, with the laying of the first foundation stone; and a large crew of stone cutters, masons and laborers worked for more than two years to complete the viaduct. The railroad had planned to use stone from a local quarry in Canton, but this proved unsuitable for other than foundation stone and backing, and granite finish stone had to be brought from a quarry on Rattlesnake Hill in nearby Sharon. This was hauled by teams of oxen or horses a distance of some 3 miles to the railroad, where it was loaded on a flat car and rolled downhill by gravity a distance of 4 miles to the construction site.

The viaduct was the final link in completing the line between the two cities. Passengers were being carried over the entire route by early June of 1835, transferring to horse-drawn carriages to get around the still incomplete structure. Trains began to cross the viaduct on July 28, 1835, making it possible for through trains to complete the journey between Boston and Providence in only two hours. Eclipsed only by Latrobe's Thomas Viaduct, completed earlier in the same month, the Canton Viaduct was indeed a prodigious achievement and the source of great regional pride. "The viaduct at Canton...," said the *Boston Advertiser,* "is a stupendous work ... it will stand for ages an enduring monument of the high talents and high attainments of its accomplished engineer."

Indeed it was all of that and more, for no one then could have visualized just how enduring and adaptable a structure McNeil had built. The Boston & Providence soon became a key link in what would become the New Haven's all-rail Shore Line route between Boston and New York, and traffic over the viaduct grew rapidly. Only 25 years after its completion the viaduct's stone parapets were removed and a heavy timber structure, projecting several feet beyond the top on either side, was installed to allow double tracking of the line. In 1885 this was replaced by an iron structure. In 1909, after one of them had begun to fail, the New Haven reinforced the arches on both sides of the viaduct with concrete. Aside from that, and a second roadway arch cut through the viaduct in 1952, William McNeil's splendid work has endured little changed for more than a century now. It was listed on the National Register of Historic Places in 1984.

Now part of Amtrak's Northeast Corridor, the remarkable viaduct is undergoing its first significant modifications in almost 90 years as the railroad gets ready for the start of high-speed train operation in 1999. The metal deck is being replaced with a wider one of precast concrete to allow the greater track spacing required for high-speed operation. The now deteriorated concrete support arches added by the New Haven in 1910 are being replaced with reinforced concrete, and 6-inch steel rods are being drilled down from the top and in from the sides at an angle to reinforce the viaduct. When the job is done, Amtrak's new American Flyer high-speed trains will operate across the viaduct at 140 mph, their speed being limited only by the 1-degree curve and not the capacity of McNeil's remarkable structure. Surely few engineering works from the dawn of American railroad history have stood the test of time so well.

A photograph of Canton Viaduct dating to about 1885 shows the wrought iron railing installed in 1880.—Canton Historical Society.

GETTING THERE

Canton Viaduct is 15 miles south of Boston at the center of Canton, which is just east of Exit 11 on I-95.

Canton Viaduct passed to still another new owner in 1976 when Amtrak acquired the Northeast Corridor from the bankrupt Penn Central. Southbound from Boston to New York, Amtrak's Northeast Direct train 172 crossed the viaduct on July 28, 1996. By this time Amtrak was planning the work that would ready the structure for the new era of high speed trains in the corridor that would begin late in 1999.—William D. Middleton.

Across the Waters

The Erie's Starrucca Viaduct (1848) at Susquehanna, Pennsylvania

If any railroad came close to the Baltimore & Ohio for the remarkable abilities and achievements of its engineers, it was the New York & Erie, which built a series of extraordinary bridges as its builders pushed the line westward across New York to Lake Erie in the 1840s. Easily the greatest of these, and certainly the most enduring, was the stone arch viaduct built to carry the Erie's main line across the valley of Starrucca Creek in northeastern Pennsylvania. Built more than a decade after such great stone bridges as the B&O's Thomas Viaduct or Canton Viaduct on the Boston & Providence, it was the last major stone structure erected in that pioneer era of railroad construction. The B&O and other lines that had built stone bridges had found them to be both a costly and lengthy effort and had turned to more quickly and cheaply built, if less enduring, wooden structures. The Erie was to learn this lesson anew.

Starrucca Creek cut across the Erie's alignment as it descended the western slope of the mountains that lay between the Delaware and Susquehanna valleys about 20 miles southeast of Binghamton, New York. More than a hundred feet deep and a quarter of a mile across, the valley was considered too deep to cross with an embankment. The Erie decided to build a viaduct instead.

Its design and construction were the work of two men. The designer was Julius W. Adams, the engineer for the railroad's Central Division, while the construction was supervised by James P. Kirkwood, his brother-in-law and an experienced engineer. Just how much of this remarkable project each of them was responsible for varies among accounts.

The structure planned for the crossing was a 1040-foot viaduct of 17 semicircular arches, each spanning 51 feet with a rise of 20 feet. The viaduct was 26 feet wide over the copings and stood 100 feet high above the level of the creek. While the line is actually on a rising northward grade of about 60 feet per mile, making the north abutment 12 feet higher than the south abutment, the viaduct was made to appear as a level structure by making the keystone and the springer course of each arch about $3/4$ inch higher than the preceding arch to the south. The structure was built largely of a local gray sandstone, called bluestone.

The foundations for two short piers on the north hillside and one on the south hillside were built of stone laid directly on the underlying hardpan, but those for 13 full height piers were built with a footing of concrete slabs. Each of these slabs was 3 feet thick, 18 feet 6 inches long, and 40 feet 3 inches wide, resting on the underlying hardpan that was found anywhere from 6 to 9 feet below the surface. The remainder of the foundation was built of stone blocks, stair-stepped up to ground level, where the foundations measured 13 feet by 33 feet 6 inches.

The viaduct's solid stone piers were on 58 foot centers and were battered in both directions, tapering from dimensions of 11 feet 4 inches by 32 feet 10 inches at the base of a full height pier to 7 feet by 25 feet at the

Starrucca Viaduct in its scenic Susquehanna Valley setting was a popular subject for such distinguished nineteenth-century artists as Jasper Francis Cropsey and Edwin Whitefield. Unfortunately, the artist for this atmospheric woodcut of the viaduct from the August 1874 *Harper's New Monthly Magazine* is not known.—Author's collection.

—American Society of Civil Engineers.

Julius Walker Adams (1812–1899) was a native of Boston. He attended the military academy at West Point during 1830–32 and then left to take up railroad engineering work with his uncle, George Washington Whistler. He came to the construction of the Erie in 1845 as a well experienced railroad engineer, having worked on the location and construction of a half dozen other lines. Although he later took on two other railroad assignments, most of Adams's long career after he left the Erie was devoted to municipal public works. He was the engineer for a number of important early water supply and sewerage systems and became a well-recognized authority in that field. He served as a colonel in the Army of the Potomac, and he was wounded in the battle of Fair Oaks, Virginia. Adams was one of the founders of the American Society of Civil Engineers in 1852, and he was its president during 1874–75. He remained active in engineering work until just a few years before his death in Brooklyn at the age of 87.

—American Society of Civil Engineers.

James Pugh Kirkwood (1807–1877) was born at Edinburgh, Scotland. After several years of study, he worked in his father's store for two years before joining a firm of land surveyors and civil engineers. In 1832 Kirkwood emigrated to the United States, where he worked on a number of railroad location and construction projects. On one of these, the construction of the mountain division of the Western Railroad of Massachusetts, he gained the experience in the building of stone bridges that served him so well at Starrucca. Following completion of the Erie construction, Kirkwood served for a time as the railroad's general superintendent and in several more railroad assignments. The remainder of his career was largely devoted to engineering work for municipal water supply systems. He was a consultant on New York's Croton Aqueduct and water projects for many other cities, remaining active in the field until the time of his death in Brooklyn at the age of 70. Like his long-time colleague Julius Adams, he was a founder of the American Society of Civil Engineers and was its president during 1867–68.

level of the springer course. The top of each pier carried a solid haunch that extended 10 feet above the spring line, above which a 4-foot-thick headwall rose to the level of the flagstone deck of the viaduct. The interiors of the arches were of open construction, with three longitudinal brick walls or relieving arches set between the spandrel walls to carry the loads from the deck to the arches. The viaduct deck was built of bluestone slabs.

The Erie had begun work on the viaduct on August 7, 1847; and by the end of the construction season, several quarries had been opened, a track laid to the principal quarry 4 miles up Starrucca Creek, and the concrete base slabs completed for six piers. But, dissatisfied with progress, Adams dismissed the contractor and brought in James Kirkwood to run the job directly, employing small contractors to cut stone and lay masonry. Intent on getting the job done before the end of 1848, the railroad had as many as 800 men at work on the site at one time.

Stone was hauled in by quarry railroad from the Stevens Point quarry up Starrucca Creek and by wagon from a Susquehanna Valley quarry, unloaded at a stone yard for cutting, and then reloaded and hauled to the viaduct site where derricks lifted it into place for laying.

The arches were erected from a timber falsework of trestles, scaffolding and arch centers. In order to speed the work, full sets of centering were erected for each arch, rather than reusing the centering as had originally been planned.

In just seven months' time during 1848, almost 18,000 cubic yards of masonry were laid, and the last arch was closed on October 10, 1848. The structure was ready for track a month later and all work was completed on November 23. The first locomotive, the Erie's famous Norris-built 4-4-0, *Orange,* crossed the viaduct on December 9, and before the end of the year, the Erie was operating regular service all the way from the Hudson River to Binghamton.

Starrucca Viaduct was a handsome structure that was "admitted to be the greatest work of art along the railroad." It had also been an extremely expensive work for the Erie's British financiers, having cost somewhere around $335,000, enough to make it the most expensive bridge ever built anywhere in the world up to that time.

Still, Starrucca proved to be a good investment that has served the Erie and its successors exceptionally well for a century and a half now, having easily assumed loads and traffic far beyond those for which it was designed. Originally built with a single track laid to the Erie's 6-foot broad gauge, the structure was able later to accommodate a second broad-gauge track, although the rails had to be gantleted. Two standard gauge tracks laid side by side replaced this arrangement in the 1880s. The Erie's great viaduct was designated a National Historic Civil Engineering Landmark by the American Society of Civil Engineers in 1973 and listed on the National Register of Historic Places in 1975. It continues to carry traffic for Conrail and the New York & Susquehanna today.

GETTING THERE

Starrucca Viaduct is located in the center of the town of Lanesboro, in the northeastern corner of Pennsylvania. Lanesboro is about 9 miles east from Exit 68 on I-81 via State Route 171 through Susquehanna.

The Baltimore & Ohio's Bollman Truss Bridge (1850)

The Baltimore & Ohio, as with so many advances in railroad engineering, was a pioneer in the use of iron truss bridges. One early design that found favor was developed by the railroad's master of road Wendel Bollman, a largely self-taught engineer. Bollman's design was sometimes called a suspension truss, a concept derived from a trussed beam. Horizontal top chord members of the truss, the vertical posts or struts which subdivided the panels, and the end posts, all of which were in compression, were cast iron sections of hexagonal cross section. A system of suspension or truss rods radiating from the top of the end posts to the bottom of the posts at each panel point, which were under tension loads, were made of wrought iron. Diagonal wrought iron rods were also installed across each panel. The floor system was of timber.

The first B&O bridge to employ the Bollman truss was a 76–foot span erected in 1850 over the Little Patuxent River on the Washington branch near Laurel, Maryland. A year later a much larger Bollman truss, spanning 124 feet, was completed at Harpers Ferry, Virginia, replacing a wooden structure on the B&O's Winchester branch. Early the following year Bollman

One of the Baltimore & Ohio's largest Bollman truss bridges was erected for its crossing of the Potomac River at Harpers Ferry, West Virginia, seen here from the hills west of the river. The Bollman trusses continued to carry railroad traffic until 1894.—Smithsonian Institution (B&O Neg. No. 690).

This three-span Bollman deck truss bridge carried the Baltimore & Ohio over the Potomac at North Branch, Maryland. The photograph shows the railroad's celebrated Artists Excursion of 1858, which conveyed more than 50 artists and photographers over the line from Baltimore to Wheeling, Virginia, making frequent stops so that the passengers could sketch or photograph the scenery.—Smithsonian Institution.

The only intact surviving example of Wendel Bollman's suspension truss bridge design is this crossing of the Little Patuxent River at Savage, Maryland. The bridge served a B&O branch line until 1947, and is now the centerpiece of a Howard County park.—William D. Middleton.

—Smithsonian Institution.

Wendel Bollman (1814–1884) was associated with the B&O from its very beginning. Bollman's formal education ended at the age of 11, after which he was apprenticed to an apothecary before taking up the carpentry trade. Then only 14 years old, he was said to have marched in the parade to the laying of the B&O cornerstone on July 4, 1828. A year later he was working as a carpenter laying the railroad's first track. Bollman left the B&O after several years to return to the carpentry trade, but he rejoined the railroad in 1838 to rebuild the timber bridge at Harpers Ferry. His practical experience reinforced by intensive self-study, Bollman rose rapidly through the B&O's engineering ranks. He was soon made foreman of bridges and then went on to become an assistant for bridge design to chief engineer Benjamin Latrobe. In 1848 Latrobe named Bollman the B&O's master of road, responsible for all of the railroad's fixed property.

In 1858 Bollman left the B&O to form his own bridge-building firm in Baltimore. The firm then contracted with the B&O to fabricate many of its bridges and filled orders for railroads as far away as Chile, Mexico and Cuba. After the Civil War, Bollman formed a new firm, the Patapsco Bridge & Iron Works, still building a number of bridges for the B&O. In addition to his patented truss, Bollman developed a design for an easily assembled wrought iron column section that was greatly superior to the cast iron sections then in general use. Bollman continued in the bridge and iron business at Baltimore until his death there at the age of 70.

obtained a patent for his design, which he called "a suspension and trussed bridge."

As one of the very first successful iron truss designs, the Bollman truss helped to advance the use of iron bridges generally, even if most railroads used other designs. The B&O itself was a principal user of the Bollman truss, and more than a hundred were built over the next two decades to replace the railroad's earlier wooden bridges. Most were located on the B&O's lines east of Harpers Ferry. Over a period of several years during and after the Civil War, Bollman trusses with span lengths of 160 feet replaced the timber trusses on the Potomac River crossing at Harpers Ferry, which had been burned by Confederate forces. The longest bridge to use the Bollman truss was the B&O's Ohio River crossing between Benwood, West Virginia, and Bellaire, Ohio, completed during 1868–71. The mile-and-a-half-long crossing had 14 spans, divided between Bollman and Whipple trusses. Bollman truss bridges on other railroads included a major span on the Mississippi River at Quincy, Illinois, for the Burlington. Completed during 1867–68, the bridge had a 360-foot central swing span made up of two Bollman trusses supported at their outer ends by chains suspended from a center tower.

New materials and better truss designs finally made the Bollman design obsolete, and none were built after

1875. Bollman's sturdy trusses, however, continued to carry traffic for many more years. The B&O's Bollman trusses at Harpers Ferry continued to carry main line traffic until 1894, and a section of the bridge remained in service for road traffic until it was swept away by a flood in 1936. Some Bollman truss bridges were still in service on the Valley Railroad of Virginia, a B&O subsidiary, until as late as 1923, and some lasted even longer on isolated branches and sidings.

The most durable of them all proved to be a bridge of two 81-foot spans that carried a B&O branch across the Little Patuxent River at Savage, Maryland, only a few miles from the site of the very first Bollman truss bridge. Originally erected elsewhere on the B&O main line in 1869, the bridge was relocated to the Little Patuxent crossing in 1887. The structure continued to carry railroad traffic until the mill it served was closed and the spur abandoned in 1947. The bridge itself, however, has remained in place ever since. Designated a National Historic Civil Engineering Landmark by the American Society of Civil Engineers in 1966 and listed on the National Register of Historic Places in 1972, the bridge was restored in 1984 as the centerpiece of a Howard County park along the river.

An October 25, 1858, view of Victoria Bridge construction by Montreal photographer William Notman shows progress in laying the stone piers and erection of the wrought iron tubes.—Notman Photographic Archives, McCord Museum of Canadian History (Neg. No. 7525).

GETTING THERE

Savage is about midway between Washington and Baltimore, between U.S. 1 and I-95. The bridge and the historic Savage Mill are just a few blocks west of downtown Savage.

The Grand Trunk's Victoria Bridge (1859) at Montreal, Quebec

In the mid-nineteenth century the broad St. Lawrence River at Montreal presented a major barrier between east and west for Canada's growing rail system. During the navigation season, ferries and barges moved passengers and freight across the river, while sleighs were pressed into service during the long winter, when the St. Lawrence was solidly frozen over. Twice every year all traffic came to a halt for several weeks as the ice began to form in the early winter, or to break up in the spring.

The idea of a bridge at Montreal was being advanced as early as the mid-1840s, and several studies all suggested as the best site a location between Point St. Charles, on the Montreal side, and St. Lambert. While this site would require a structure nearly 2 miles long, it was upstream from the harbor used by tall-masted ocean-going vessels, and the river was quite shallow over much of the crossing. The main channel used by steamers was only 300 feet wide and was about 25 feet deep at the summer low water level.

Alexander Ross, chief engineer of a British firm that was assisting the Canadian government in railway development, looked over the site in 1852 and proposed the construction of a tubular iron bridge. Ross had long been associated with the great British railway engineer Robert Stephenson, who came to Montreal in 1853 to view the planned site and subsequently joined with Ross to design the structure.

The bridge reflected the experience of its British engineers and was unlike anything else being built in North America. A few years earlier Stephenson had completed two innovative tubular bridges in Wales for the Chester & Holyhead. Instead of the iron trusses being used by American engineers, both bridges employed a huge wrought iron box girder as the principal structural element, with trains passing through the tubular structure.

In overall size, the Montreal bridge dwarfed Stephenson's earlier works. Stone piers supported 24 tubular iron spans of 242 to 247 feet that rose gradually from either end to a 330-foot center span over the navigation channel, which provided a clearance of 60 feet above the water level. This created a continuous tubular structure 6592 feet long, while the overall length of the bridge, including abutments and approaches, was 9144 feet.

The bridge was built by a British firm, Peto, Brassey & Betts, which brought many of the skilled construction workers it would need from Great Britain. Construction work was headed by British engineer James Hodges. At the peak of activity during the five-year construction period, Hodges had close to 3000 men at work on the bridge.

Erection of the bridge's 330-foot center tube was supported by a timber Howe truss and cribbing placed on the river ice. William Notman photographed the work in March 1859 as the builders raced against time, completing the work just a day before the ice went out on March 28.—Notman Photographic Archives, McCord Museum of Canadian History (Neg. No. 7534).

Construction of the abutments and piers began in the spring of 1854. After the ice had gone out and the river level dropped, floating wooden cofferdams that had been fabricated during the winter were floated into position and sunk into place for building the first two piers. These were then pumped out and the stone piers built within them on the river bed. The floating cofferdams proved difficult to seal against the uneven bottom, and those for most of the remaining piers were built with wooden cribbing which was laid in the river and sunk into place by the weight of stones and subsequent layers of cribbing. The two massive stone abutments were built in a similar manner, while fill material, faced with stone rip-rap, was placed for the long approaches to the bridge.

It was difficult work. Strong currents hindered placement of either floating cofferdams or cribbing, and it was hard to get a watertight seal for the cofferdams on the irregular river bottom. The swift current created hazards to the work. Barges floating down river often collided with the structure, the contractor's barges were sometimes overturned, and most of the 26 workers killed on the project lost their lives to drowning. Every fall, the annual freeze-up brought construction to a halt.

Despite all the problems, the bridge gradually took shape. More than 2.7 million cubic feet of limestone for the foundations was hauled in from Quebec and Vermont quarries. Tugs and barges hauled the cut stone out into the river for placement in the piers and abutments.

The foundation work was well advanced by the end of the 1856 season, and erection of the rectangular tubes began the following year. These were made of wrought iron plates riveted together with "T" sections. Each tube was 16 feet wide and 18 feet 6 inches high at the ends, increasing to 22 feet at the center of each span. All of the iron work was fabricated at the contractor's Canada Works in Birkenhead, England, where almost 5000 pieces for each tube were cut to shape, punched and marked for assembly. These were then shipped to Canada in 25 ships, one for each tube.

Most of the tubular spans were erected from temporary scaffolding supported by timber cribs sunk into place between the piers. The first tube was completed late in 1857, and by the winter of 1858-59 Hodges was ready to begin erection of the long center span. A temporary timber truss, spanning between the two piers and supported by intermediate cribbing placed on the ice, provided a platform for the erection work. This began at the end of January and proceeded day and night through the bitter winter weather until the tube was

"Grand Finale of Fire-Works in Honor of the Prince of Wales and the Successful Completion of the Victoria Bridge," an engraving by G. A. Lilliendahl of New York in the September 1, 1860, *Harper's Weekly,* depicted the final event in the celebration of the completion of Montreal's great bridge. —Author's collection.

Across the Waters   23

An early general view shows Victoria Bridge in its original tubular form. Conventional steel trusses took the place of the novel iron tubes in the reconstruction of 1897–98.—Smithsonian Institution (Neg. No. 92-15119).

complete eight weeks later, just a day before the ice went out.

Late that year all remaining work had been completed and the first train crossed the span on December 12, 1859. A ceremonial opening came the following summer as the climax of a royal visit to Canada. On August 25, the Prince of Wales laid the final stone in the north abutment with a silver trowel and then proceeded to the center of the bridge to drive the final rivet in what was to be called Victoria Bridge. A few nights later a spectacular fireworks display completed the opening festivities.

Robert Stephenson's innovative tubular bridge design was never repeated in North America, but his sturdy Victoria Bridge gave long service to the Grand Trunk. By the end of the century, however, increasing traffic and heavier trains had made Stephenson's single-track structure obsolete. During 1897–98 the stone piers were widened and new steel truss spans were erected around the tubular sections, which were then removed. The new

The steel truss bridge of 1897–98 was extensively altered in 1959 to add lift spans and bypass routes for rail and road traffic at the St. Lambert Lock of the new St. Lawrence Seaway. En route from Montreal to Quebec, VIA Rail's train 22 crossed the much-modified structure in July 1998.—William D. Middleton.

Robert Stephenson (1803–1859) and his father George Stephenson were towering figures in the formative years of British railways. Born at Willington Quay, near Newcastle, Robert was educated at a school in nearby Newcastle. He was apprenticed for three years to one of his father's associates at Killingsworth Colliery, and he then attended Edinburgh University for nearly a year. Robert worked with his father on surveys for two pioneer railways and in locomotive development, culminating in the triumph of the Stephenson *Rocket* at the famous Rainhill trials of 1829.

The younger Stephenson developed a wide reputation for his proficiency in every aspect of railway engineering, He became particularly skilled in bridge engineering. He completed cast iron and stone arch bridges at Newcastle and Berwick in 1849 and 1850, while his innovative tubular iron bridges for the Chester & Holyhead were opened in 1848 and 1850. Stephenson built two more of his tubular bridges for crossings of the Nile on Egypt's Alexandria & Cairo Railway, but the last and greatest of them all would be Victoria Bridge at Montreal.

In addition to his engineering work, Stephenson served in Parliament from 1847 until his death. He died at London, just two months before the first train crossed Victoria Bridge, and was buried in the nave of Westminster Abbey.

Robert Stephenson.—Science Museum, London.

Alexander M. Ross, from The Illustrated London News of February 19, 1859.—Library of Congress.

Alexander McKenzie Ross (1805–1862) was born in Scottsburn, Ross-shire, but later moved with his family to Dornoch in Sutherlandshire, where he excelled in the study of mathematics in the local public school. He was employed at the age of 18 by a London public works contractor and later became a railway manager on the North Midland. In 1843 Ross joined Robert Stephenson on the Chester & Holyhead, where he assisted in the construction of the Conway and Britannia tubular bridges. He went to Canada in 1852 to work in railway development. There he became chief engineer for the building of the Grand Trunk and of Victoria Bridge. He died at Montreal just two years after the triumphal opening of the bridge.

Victoria Jubilee Bridge provided space for two tracks, as well as road traffic.

Still more change came in 1959 with the completion of an elaborate bypass at the south end of the bridge for the new St. Lawrence Seaway's St. Lambert Lock. One lift span was constructed in the bridge at the downstream end of the lock, while a second at the upstream end was linked to the bridge by a detour structure. Thus, as water traffic moved through the lock, one lift span or the other was always available for rail traffic. Today, though much modified over its 140 years, the great bridge continues to provide a vital link in Canadian National's transcontinental network.

GETTING THERE

Victoria Bridge crosses the St. Lawrence between the Pointe-St.-Charles district in Montreal and St. Lambert, and is crossed by Rue Bridge (Route 112).

"The St. Louis Bridge during Construction," an engraving from Scribner's *The American Railway* (1892), shows how the arches were constructed by cantilevering them symmetrically outward from the completed piers, supported by cables from temporary support towers until they were joined. This enabled the bridge to be erected without obstructing navigation with temporary falsework.—Author's collection.

The St. Louis Bridge (1874) at St. Louis, Missouri

Steamboat interests centered at St. Louis had bitterly opposed the building of the first railroad bridge across the Mississippi at Rock Island, Illinois, in 1852. But scarcely a decade later, as St. Louis saw the new railroad center at Chicago gain trade at its expense, the city was eager for a bridge of its own. A bridge company was formed in 1865, and charters were obtained from the Missouri and Illinois legislatures. The following year Congress passed an authorizing act as well, and the company was soon ready to begin construction.

James Buchanan Eads, an innovative, self-taught engineer, was appointed chief engineer for the bridge. It was a curious choice, for Eads had never built a bridge, but he went on to build one of the most important bridges of the nineteenth century.

Eads proposed to span the river with three great steel arches. Steel was much stronger than cast or wrought iron, and the new Bessemer process was just beginning to make it both plentiful and cheap. Its adoption for the main arch spans of the bridge would be the key to Eads's success.

Building the foundations for the bridge was a difficult task. With a healthy respect for the way the swift Mississippi currents could scour the material on the river bottom, Eads insisted that the foundations be founded on the bedrock that lay below the river bottom. At the site selected for the bridge, opposite the foot of Washington Avenue at the heart of the St. Louis riverfront, this was anywhere from about 50 feet below the high water mark on the St. Louis side to nearly 130 feet on the Illinois side. To get to bedrock Eads would have to sink his foundations below some 25 feet of water and as much as 80 feet of sand.

Eads began work on the west abutment in August 1867. The bedrock here was only 47 feet below high water, and he built the foundation with a conventional open cofferdam of wood. But during a visit to Europe early in 1869, Eads studied the use of pneumatic caissons, and he returned to St. Louis determined to employ that method for the remaining foundations, which had to go much deeper. It was the first major use of this method anywhere in North America, and the caissons required were larger and were taken to greater depths than any before them.

The first of these was built in the spring of 1869. This was a 437-ton timber caisson for the east river pier that was 82 feet long, 60 feet wide and 9 feet high. The outside surface was covered with riveted iron plating, and the top was reinforced with iron girders. Cylindrical iron air locks projected into work chambers inside the caisson. An iron cutting edge cut through the river bed as the caisson descended toward bedrock.

The caisson was launched in October 1869, towed into position at the bridge site, and anchored in place by piles driven into the river bottom. Air was pumped into the caisson, and workmen began the task of building the masonry pier on its top as it gradually sank into the river under the weight of the masonry. Ten 25-foot-long suspension screws guided the descent.

Once the cutting edge was completely embedded in the sand of the river bottom, work began inside the caisson work chamber. Workers descended a spiral stairwell in the center of the pier and passed through an air lock to begin work inside the pressurized chamber. About 30 men at a time worked in the chamber, shovel-

St. Louisans were intensely proud of their great new steel arch bridge across the Mississippi. This handsome engraving of the St. Louis Bridge dates to about 1880. Trains crossed the bridge on a double-track line on the lower level that was linked to the city's Union Depot by a mile-long tunnel under the downtown. Vehicles, streetcar lines, and pedestrians used the upper level.—Author's collection.

ing the sand from the river bottom into a water-powered sand pump devised by Eads. Work continued around the clock, and the caisson settled toward bedrock at a rate of as much as 20 inches a day.

Work began with a slightly smaller caisson for the west pier in January 1870, and with a third caisson for the east abutment on the Illinois side the following summer. Here, where the rock was deepest, the caisson was sunk to a depth of 136 feet below high water level.

Construction of the piers and abutments was fraught with problems. Ice floes and a tornado wrecked the superstructure of the east abutment, and the east pier was damaged by ice and flooding. The most serious problem, however, was the little known or understood affliction of caisson workers known as "caisson disease," or simply the "bends," which was encountered as the men worked at unprecedented depths and pressures. Nearly a hundred men were stricken and 14 of them died. The cause was finally attributed to too rapid decompression from the high air pressures in the deep caissons, and Eads devised a decompression chamber that greatly reduced the danger to the workers.

Work continued even as Eads sought a solution to the problem. The east pier reached bedrock at the end of February 1870, and the west pier followed several months later. The last and deepest caisson at the east abutment reached bedrock in March 1871.

Eads's design for the bridge structure incorporated two arches of 502 feet and a central arch of 520 feet. These supported a double-track railroad on a lower deck and a roadway on an upper deck. Four pairs of steel arches, their upper and lower members set 12 feet apart and laced together with diagonal bracing, made up each span. The arches were fabricated in tubular sections 18 inches in diameter and 12 feet long. At first Eads's specifications for the steel were impossible to meet, but after months of effort, supplier Andrew Carnegie was finally able to satisfy his strict requirements.

The erection work began in April 1873. To avoid obstructing the river channel with falsework, Eads erected the arches by cantilevering them outward symmetrically from the piers. Cables from temporary sup-

Well into its second century, Eads Bridge remained a graceful element of the St. Louis riverfront at the time of this April 1987 photograph. The trains had long since ceased running over the span, but within a few years it would be back in business as a key link in the region's new MetroLink light rail system.—William D. Middleton.

Across the Waters   27

port towers on each pier supported the half-arches until they were joined at the midpoint and became self-supporting. The joining of the first arch brought new problems. Eads had planned to connect the two sections by inserting connecting tubes, but the metal expanded so much in the hot weather that the tubes could not be inserted. Tons of ice applied to the arch sections failed to reduce the expansion sufficiently, and the problem was finally solved through the use of special adjustable closing members designed by Eads.

The last arch was closed on December 18, 1873, and the bridge was complete late the following spring. On July 2, 1874, the safety of the completed bridge was demonstrated to the public as 14 locomotives were coupled together, seven on each track, and moved back and forth across the structure. The first railroad crossing of the lower Mississippi was dedicated with elaborate festivities on the Fourth of July.

Although it was an engineering triumph, the bridge was not an immediate financial success. The bridge company was bankrupt within a year, and the bridge was sold at public auction a few years later for less than a third of what it had cost to build it. But in time, the bridge helped assure St. Louis an important place in the new transcontinental railroad system.

After a century of service, inadequate clearances idled the structure's rail deck in 1974, and the bridge faced an uncertain future. Then it turned out to be just what the region needed for a new light rail line across the river. Some repairs were made, new tracks were laid, and in July 1993 MetroLink trains began using the structure. James Eads's St. Louis Bridge could look forward to a second century as productive as its first.

Now known as Eads Bridge in honor of its builder, the bridge was listed on the National Register of Historic Places in 1966 and was designated a National Historic Civil Engineering Landmark by the American Society of Civil Engineers in 1971.

GETTING THERE

Eads Bridge crosses the Mississippi to East St. Louis from the foot of Washington Avenue in downtown St. Louis. Good views are available from the park surrounding the Gateway Arch, just to the south on the St. Louis side, or from the pedestrian walkways on the top deck of the bridge. MetroLink light rail trains cross the bridge on the lower level.

—Library of Congress.

James Buchanan Eads (1820–1887) was born at Lawrenceburg, Indiana. Although his formal schooling ended at 13, the self-taught Eads became one of the great American engineers of his time. At the age of 18 he began a lifelong association with the Mississippi River as a steamboat clerk. Only a few years later Eads designed an innovative diving bell vessel for salvage work on the river, and he soon became wealthy in the salvage business. During the Civil War, he designed and built a fleet of ironclad gunboats that helped the Union forces to gain control of the Mississippi and its tributaries.

After he completed the St. Louis Bridge project, Eads planned and built the system of jetties that assured access to the Mississippi from the Gulf of Mexico. Later he completed still other waterways projects all over the United States, and in Mexico, Canada and England. In 1880 he began work on the ambitious Tehuantepec Ship Railroad in Mexico that would have carried ships between the Gulf of Mexico and the Pacific. In 1884 he became the first American ever awarded the Albert Medal of the British Society for the Encouragement of Art, Manufactures and Commerce. He died at Nassau, Bahamas, while visiting there to recover from illness.

The Cincinnati Southern's Kentucky River Bridge (1877) at High Bridge, Kentucky

From an early date, Cincinnati had sought a rail route across Kentucky to the south. A formidable barrier to such a route, however, lay scarcely a hundred miles south of Cincinnati, where the Kentucky River flowed through a deep gorge in the plateau south of Lexington.

A first attempt to bridge the gorge was launched in 1853, when the new Lexington & Danville contracted with bridge engineer John A. Roebling to design and build a suspension bridge at a point just below the confluence of the Dix and Kentucky rivers. Roebling, who was then building his great suspension bridge over the Niagara River (see page 175), designed an even longer bridge with a 1224-foot main span for the Kentucky River crossing. Construction began in 1854, but the work was only partially complete when the financial panic of 1857 brought work on both the bridge and the railroad to an end.

The project came alive again almost 20 years later. By this time the City of Cincinnati had organized the Cincinnati Southern to build its own line south across Kentucky. This new company acquired the properties of the ill-fated Lexington & Danville and set out to bridge the Kentucky at the same site Roebling had chosen.

The side spans for the Kentucky River High Bridge were erected by cantilevering them outward from the bluffs to a temporary tower at mid-span. They were then cantilevered the remaining distance to permanent intermediate iron bents. In this November 14, 1876, photograph by Lexington photographer James Mullen, the erection of Span No. 1 has been completed almost all the way to the intermediate bent. Once the two side spans were complete, the center span over the river was erected by cantilevering its two halves outward from the intermediate bents, meeting at mid-span without any need for temporary scaffolding.—Smithsonian Institution (Neg. 82-1439).

The bridge that took shape over the gorge during the next several years was the work of two of the leading bridge engineers of the time, and it would represent some important advances in bridge engineering. Louis G. F. Bouscaren, the principal assistant engineer of construction for the Cincinnati Southern, was one of the pioneers in the use of comprehensive bridge specifications as the basis for competition among bridge design and construction firms. His specifications for both the Cincinnati Southern's crossing of the Ohio at Cincinnati and the Kentucky River bridge were considered models of good practice. Both incorporated an improved system of concentrated loads for bridge design and required load testing of full-sized structural members. C. Shaler Smith was president and chief engineer of the Baltimore Bridge Company, which was awarded a contract for the design and construction of the bridge in July 1875, and he would be its designer.

High Bridge, as it came to be called, would be the

Visible in this photograph of a train on the High Bridge are the two stone towers that survived from John A. Roebling's abandoned suspension bridge project of 1854–57 at the site. The towers were used by Charles Shaler Smith in the erection of the bridge during 1876–77, and remained in place until 1929, when they were removed to permit double-tracking of the line over the bridge.—Library of Congress (Neg. LC-D4-19984).

Across the Waters

Replacement of the High Bridge with a heavier structure during 1910–11 was an extraordinarily difficult task. This construction photograph shows how the new, larger and heavier trusses were erected around the original structure, which remained in service throughout the work.—Smithsonian Institution.

highest bridge in the world, if only for a few years; and it would be the first significant bridge in North America to employ the cantilever truss form. The bridge would be notable in another way, too, for it would be one of the last major North American bridges built entirely of wrought iron; steel would entirely displace wrought iron as a bridge material over the next two decades.

An important reason for the adoption of the cantilever form was the need to minimize the use of falsework in the 275-foot- deep gorge, which was subject to severe flash floods. The bridge designed by Smith was 1138 feet long between abutments and was made up of three equal spans of 375 feet each. These were supported by two intermediate bents carried on stone piers built on either side of the river. The Whipple type trusses were 37 feet 6 inches deep and 18 feet wide, and carried a single track.

The design reversed the usual arrangement for cantilever bridges, with a center span that was continuous

The Cincinnati Southern later became an important route in the Southern Railway system. Here, three Electro-Motive F3 diesel units pull a southbound Southern freight across High Bridge.—Southern Railway.

Charles Shaler Smith (1836–1886) was born in Pittsburgh. After his formal education ended at an early age upon the death of his mother, Smith took up work as a surveyor for the Mine Hill & Schuylkill Haven Railroad, which led to a notable career in railroad and bridge engineering. He joined the L&N in 1855 to begin a decade of engineering work for several railroads with time out for Civil War service as a Confederate captain of engineers.

—Smithsonian Institution.

In 1866, Smith joined with Benjamin and Charles Latrobe, the sons of Benjamin Latrobe, II, to establish the Smith, Latrobe & Company bridge building firm at Baltimore, which became the Baltimore Bridge Company three years later. In addition to the Kentucky River bridge, the firm built a number of other important bridges, notably a bridge across the Missouri River at St. Charles, Missouri; a crossing of the Mississippi River at Rock Island, Illinois; and major cantilever crossings of the Mississippi at St. Paul and of the St. Lawrence at Montreal. The Baltimore firm was disbanded in 1880, and Smith then worked as a consultant until his death at St. Louis at the age of 50.

Louis Gustave Frederic Bouscaren (1840–1904) was born on the island of Guadeloupe and came to the United States in 1850 when his family settled in Grant County, Kentucky. He was educated at home and a school in Cincinnati before studying at the Lycée St. Louis at Paris. He then entered L'Ecole Centrale des Arts et Manufactures, graduating with high honors as an engineer in 1863.

Returning to the United States, Bouscaren began a distinguished career in railroad and bridge engineering. After a number of years with the Ohio & Mississippi, he joined the Cincinnati Southern in 1873 to develop specifications and supervise construction for the railroad's crossing of the Ohio at Cincinnati. His work in preparing model specifications for what was to be the longest truss bridge in the world and his development of a new system of design loading and testing procedures for full-sized structural members did much to establish his wide reputation in bridge engineering. Shortly after completing the Kentucky River bridge, which followed close behind the Ohio River span, Bouscaren became principal engineer for the Cincinnati Southern and completed construction of the line to Chattanooga.

Over the next decade he was engaged in the engineering for several new railroads in the south central states. Bouscaren returned to Cincinnati in 1886 to establish his own consulting engineering practice. His work there included the development of the water supply works for both Cincinnati and Covington, Kentucky, and a number of important bridge projects. In 1894 President Grover Cleveland named him to a commission appointed to study and report on the best plan for bridging the North (Hudson) River at New York.

over the two intermediate bents and cantilevered beyond them on either side. Two 300-foot "semi-suspended" simple trusses, each with one end supported at the abutment and the other hinged to the cantilever arm of the center span, completed the structure.

Smith had worked with James Eads on the St. Louis Bridge, and he employed a similar cantilevering method to construct the Kentucky bridge. Using the stone towers and anchorages of the abandoned Roebling project, Smith erected the side trusses by cantilevering them outward from the bluffs at each side to their midpoint, where they were supported by temporary towers of wooden scaffolding. The remainder of each side span was then cantilevered from these temporary supports to the intermediate iron bents. Finally, the two halves of the center span were cantilevered from the bents, meeting exactly in midair at the center without a need for any temporary scaffolding at all in the main span over the river.

At its completion the bridge was actually a single continuous truss spanning from bluff to bluff, and it was then converted to its final cantilever arrangement by cutting the bottom chords of the side spans at the proper point and converting them into sliding joints to provide the desired cantilever action.

Erection of the bridge began in October 1876, and the structure was completed the following spring. On April 20, 1877, the railroad successfully completed an extensive series of load tests on the completed structure, and the bridge was placed in service. Formal dedication of the bridge came a little over two years later at ceremonies attended by President Rutherford B. Hayes.

Soon after its completion, the Cincinnati Southern was leased to the new Cincinnati, New Orleans & Texas Pacific Railway, which in turn became part of the Southern Railway System in 1895. With its connections, the CNO&TP formed a through route between Cincinnati and New Orleans that has remained an important Southern, and now Norfolk Southern, route ever since.

The bridge served the railroad well for more than three decades, before increasing locomotive and train weights finally made it obsolete. During 1910–11 the

Southern replaced the structure in an extraordinary effort that was almost as difficult, in its own way, as the construction of the original bridge. This new steel span, designed to accommodate two tracks, was designed by bridge engineer Gustav Lindenthal.

To support the new span, the two stone piers were capped with concrete; and new, heavier steel supporting bents were built around the original wrought iron bents. An entirely new superstructure was then constructed around the original structure, which remained in service throughout the reconstruction. The new steel trusses were wider than the original trusses and 31 feet deeper, providing room to contain both the original bridge and trains within them as they were erected. New approaches were graded at each end to reach the higher deck of the new bridge. All was complete late in the summer of 1911, and the first train crossed over the new High Bridge on the morning of September 11.

In 1985 High Bridge was designated a National Historic Civil Engineering Landmark by the American Society of Civil Engineers.

GETTING THERE

High Bridge is about 20 miles southwest from downtown Lexington. Good views of the span are available from the north end at High Bridge, which is reached from Lexington via U.S. 68 and State Route 29 through Wilmore.

The Erie's enormous wrought iron viaduct over the valley of Kinzua Creek was notable for the speed with which it was erected. Once the foundations were ready in May 1882, erection crews put up the 1500-ton superstructure in less than four months.—Smithsonian Institution.

The Erie's Kinzua Viaduct (1882) near Mt. Jewett, Pennsylvania

At the beginning of the 1880s the New York, Lake Erie & Western, one of the Erie's several bankruptcy incarnations, extended a new line south into the rich Allegheny Mountain coalfields of northwestern Pennsylvania. The railroad's builders encountered a sizable obstacle in the valley of Kinzua Creek, about 15 miles south of Bradford, Pennsylvania. At the point where the most favorable alignment crossed the valley, the creek itself was small and shallow; but it flowed through a rugged gorge that was some 300 feet deep and close to a half-mile wide at the top.

The Erie spanned the creek with a massive wrought iron viaduct that would claim the title of the highest railway bridge in the world for at least a few years after its completion. The structure was designed by Adolphus Bonzano and Thomas C. Clarke of Clarke, Reeves & Company of Phoenixville, Pennsylvania, which contracted with the railroad to design and fabricate it.

The single track viaduct was 2051 feet long between abutments and 301 feet high above the stream bed. The track was carried on a Warren truss structure 6 feet deep and 10 feet wide. This was divided into 20 spans of 61 feet and one of 62 feet, spanning between supporting towers, alternating with another 20 spans of 38 feet 6 inches spanning across the top of each tower. The 20 supporting towers ranged in height from 16 feet to just over 278 feet. Each tower was 38 feet 6 inches long and 10 feet wide at the top. The tower columns spread outward at a batter of 2 inches per foot, making the highest towers 103 feet wide at the base. Foundations for the towers were built of sandstone quarried at the site.

Aside from its great size and height, Kinzua Viaduct was notable for the speed with which it was erected. Work began at the site in August 1881, and the foundations were ready for erection to commence the following May. This was begun on May 10 and over 1500 tons of iron was erected in less than four months by a force averaging about 125 men. The work was accomplished without the use of any temporary scaffolding. The towers were erected to almost their full height through the use of gin poles, while their topmost sections and the truss spans were placed in position by a traveling steam derrick, which advanced out across the structure from the north abutment as each section was completed. The superstructure was complete on August 29, 1882, and

Erie trains were soon hauling a heavy traffic out of the railroad's rich coal lands in Jefferson and Elk counties.

The Erie's enormous iron viaduct proved short-lived. Within less than two decades the impact and vibration of increasing traffic volumes, as well as much heavier locomotives and cars, had begun to take their toll; Kinzua Viaduct was subject to increasingly severe restrictions on loading, train speeds, and brake applications. By 1900, it was time to replace the structure.

The second Kinzua Viaduct was identical in its overall dimensions to the first and stood on the same foundations. Reflecting train weights that had doubled in the intervening 20 years, the new riveted steel bridge required almost 3400 tons of metal, more than twice the weight of the first span. While the tower columns on the original structure had been built-up lattice sections about 1 foot square, the new columns were 2 feet by 3 feet. Instead of truss spans as in the older structure, the track was carried on heavy plate girders. Those spanning between towers were 6 feet 6 inches deep, while the shorter spans across the tops of the towers were 4 feet 6 inches deep. The new structure was notable for its use of

Briefly ranking as the world's highest railway bridge, Kinzua Viaduct became a popular attraction. This illustration of a train passing over the towering structure is from the May 13, 1882, *Scientific American*.—Library of Congress.

Replacement of the viaduct with a heavier structure in 1900 was accomplished with two of these 180-foot-long travelers. Beginning from opposite ends, they moved over the original structure, dismantling it and erecting the replacement viaduct as they went. —Smithsonian Institution.

The Erie's four-unit Alco freight diesel No. 730 crossed Kinzua Viaduct with a train on March 5, 1949.—W. G. Thornton, David P. Morgan Memorial Library.

viaduct towers without the usual transverse diagonal wind bracing. Instead, the design relied upon heavy brackets at the junction of the horizontal members to form a rigid structure capable of withstanding the wind loads.

The new viaduct took the place of the old under an ingenious procedure which, in effect, utilized the old viaduct as scaffolding to erect the new. Two 180-foot-long wooden truss "travelers" were built. These were long enough to rest on two supporting towers, spanning one intermediate tower, and were equipped with hoisting engines, jib cranes and an overhead track on which two traveling carriages hauled materials. The two travelers started from opposite ends of the viaduct, lifting out the old trusses, which were then placed on flat cars and hauled away, while the towers were taken apart and dropped to the ground. The new towers and plate girders were then erected, and the traveler advanced the length of another two towers and the process was repeated. After the work was complete, a derrick car lifted the old column sections up to track level for loading on cars and removal.

The Erie decided not to attempt to replace the structure under traffic. Trains were detoured and the work carried out in the shortest possible time. Dismantling of the old viaduct began May 24, 1900, and the last girder was placed on September 6. Trains were running over the viaduct again on September 25.

From the beginning, the enormous viaduct ranked as one of the wonders of western Pennsylvania, drawing a regular stream of sightseers. At one time the Erie ran Sunday excursions to the bridge from points as far away as Binghamton, New York. Erie trains quit running across the great bridge in 1957, when the railroad abandoned a section of the branch in favor of trackage rights over a parallel B&O line. Kinzua Viaduct still stands there today, however, as the centerpiece of the Kinzua Bridge State Park. Visitors still come to view what was once the highest bridge in the world; and a tourist railroad, the Knox & Kane, operates excursions to the park and over the bridge from Marienville and Kane, Pennsylvania. The structure was placed on the National Register of Historic Places in 1977 and was designated a National Historic Civil Engineering Landmark by the American Society of Civil Engineers in 1982.

GETTING THERE

Kinzua Viaduct is located in the Kinzua Bridge State Park, which is 3½ miles northeast on U.S. 6 from the town of Mount Jewett, about 16 miles south of Bradford in northwestern Pennsylvania.

The Great Northern's Stone Arch Bridge (1883) at Minneapolis, Minnesota

Long before the mid-nineteenth century, North American railroads had turned away from the use of stone masonry bridges and viaducts in favor of wood—and later iron—structures. But during 1881–83 a remarkable exception to this direction in bridge engineering took shape just below St. Anthony Falls in the Mississippi River at Minneapolis.

After more than two years of work that saw nearly 49,000 cubic yards of masonry placed in the structure, stonemasons were nearing the end of their work on the Stone Arch Bridge when this construction view was taken sometime in 1883.—Burlington Northern Santa Fe.

In a classic scene from the steam era, a Great Northern photographer recorded the railroad's premiere train, the *Empire Builder,* on the Stone Arch Bridge against a backdrop of Minneapolis flour mills. Class P-2 4-8-2 Mountain type locomotive No. 2517 headed the westbound train.—Burlington Northern Santa Fe.

James J. Hill, the Great Northern's legendary "empire builder," brought this remarkable structure into being. Hill had seen a need at Minneapolis for both a new crossing of the Mississippi and a union station, and he picked a location at the falls for a bridge that would lead to a central station site in downtown Minneapolis. In 1881, Hill provided the necessary financial backing to the Minneapolis Union Railway, which built the new connecting lines in the city, the union station, and the bridge.

The viaduct that took shape under the direction of the Union Railway's chief engineer, Colonel Charles C. Smith, reflected the difficult nature of its site. While an initial plan had called for an iron bridge to be located above the falls, Smith persuaded Hill to build the bridge of stone and to locate it below the falls. Stone would be more durable, he argued, while a site below the falls offered better conditions for constructing the piers and less risk of damage from ice. And since the 18-foot-high falls marked the upper limit of navigation in the river, the arches of a stone bridge would not impede navigation.

The 2100-foot-long structure designed by Smith crossed the river at an angle and then curved to the north as it approached the west bank. A 1283-foot section of the viaduct was built on a tangent, while an 817-foot section at the west end was constructed on a 6-degree curve. The structure had an overall height of 82 feet and stood 65 feet above water level. A 24-foot width provided space for a double-track line. The viaduct had a total of 23 semicircular arches, with four spans of 100 feet, 16 of 80 feet, and three of 40 feet.

Work began in February 1881, and the railroad soon learned what a costly, time-consuming project a stone masonry structure could be. Despite an enormous work force, it would be almost three years before the work was completed. In February 1882, for example, *The Minneapolis Tribune* reported that the contractor had a daytime force of 400 men engaged in the work, with another 100 men working at night. Hundreds more were at work in the quarries that produced the 100,000 tons of stone required for the structure.

The piers near each end of the bridge were founded on a limestone ledge which was continuous with the ledge in the bottom of the river above the falls. Caissons

Every railroad that reached the Great Northern's Minneapolis station from the east used the Stone Arch Bridge to get there. Handsome in a livery of apple green and English stagecoach yellow, the Chicago & North Western's eastbound train 400, the *Twin Cities 400,* crossed over the bridge as it began its high speed journey to Chicago on March 15, 1959.—William D. Middleton.

were sunk 6 to 8 feet below the bottom of the river to found the intermediate piers on a solid sandstone below the limestone ledge. The pier footings were made of a Portland cement concrete.

Once the piers were in place, wooden centers were erected for construction of the arches. The piers above the level of the footings were made of granite backed with limestone. The ring stones, arch sheeting, spandrel walls, and coping were made of a limestone from quarries in Minnesota, Wisconsin, and Iowa. Work continued on the structure through the winter of 1882–83, and the water used in the mortar had to be heated during freezing weather. All told, nearly 49,000 cubic yards of masonry went into the structure.

The only stone arch bridge ever built across the Mississippi, the structure was much admired, and its completion in November 1883 was the occasion for a great municipal celebration. The Stone Arch Bridge soon became a widely recognized symbol of the city and of the Great Northern, second only, perhaps, to the railroad's famed mountain goat emblem. "It is constructed for a thousand years," said an 1883 report of the Minnesota Railroad Commission, "and with its massive masonry and lofty arches recalls the solidity of the work of Roman antiquity."

Jim Hill's stone arch bridge served Minneapolis and its railroads well for the next 99 years. During 1907-11 a new drainage system was installed, and the structure was reinforced with steel tie rods and reinforced concrete arch rings placed on top of the original stone arches. In 1925 portions of the stone copings at the top of the spandrel walls were cut away to permit a wider track spacing for larger locomotives.

More changes were made in 1962 in a U.S. Army Corps of Engineers project that did major aesthetic damage to the graceful structure. To extend navigation above St. Anthony Falls, the Corps built a lock for barge traffic on the west bank of the river. Just at the point where the bridge began its graceful curve toward the west bank, a pier and two arches were removed and replaced with an ungainly 200-foot-long steel truss span to provide a clear channel for the lock entrance.

Still more work on the bridge came unexpectedly in the wake of record Mississippi River floods in the spring of 1965, when three piers were severely weakened by flood waters, causing a 100-foot section of the bridge to settle. One pier had dropped more than a foot. The bridge was closed for six months while emergency repairs were made. When the job was done the tracks had been brought back up to level, but there was a permanent dip in the stonework of the viaduct.

The years of railroad service came to an end for the Stone Arch Bridge after Amtrak shifted its trains to a new route through the Twin Cities in 1981, and the last Burlington Northern train crossed the structure in 1982. Although the trains are gone, Jim Hill's splendid bridge remains a permanent fixture of the Minneapolis riverfront. Rehabilitated in 1994 as a pedestrian and bicycle trail and a key segment of the St. Anthony Falls Heritage Trail, the bridge affords an incomparable view of the Mississippi River and St. Anthony Falls. The bridge is included within the St. Anthony Falls Historic District, which was listed in the National Register of Historic Places in 1971. The bridge was designated a National Historic Civil Engineering Landmark by the American Society of Civil Engineers in 1976.

GETTING THERE

The Stone Arch Bridge is located just below St. Anthony Falls in downtown Minneapolis, and has been rehabilitated for pedestrian and bicycle use as part of the St. Anthony Falls Heritage Trail. Maps and signs are available for self-guided tours of the trail, and guided tours are available in season from a program office at 125 Main Street, S.E.

The Hudson River Bridge (1888) at Poughkeepsie, New York

A bold venture of the new era of steel bridge construction that began in the 1870s was the construction of an enormous span, more than a mile in overall length, across the Hudson River at Poughkeepsie, New York. The driving force behind the project was the need for an

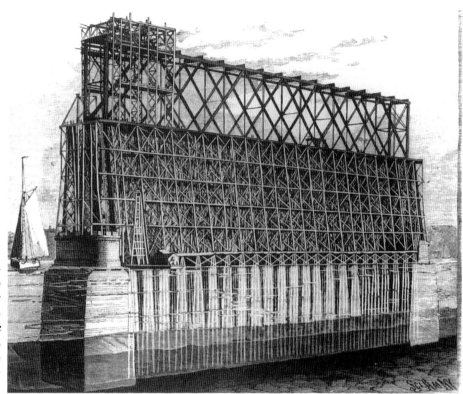

The difficult work of building the unprecedented cantilever crossing of the Hudson at Poughkeepsie was widely followed by the popular press of the time. This drawing from *Scientific American* for February 4, 1888, shows how the main anchor trusses were erected on falsework.—Author's collection.

improved rail link between New England and the rest of the country. The Hudson had not yet been bridged anywhere in the 150 miles below Albany, and there was no direct rail route to Pennsylvania or the south.

Backed by Poughkeepsie business interests who wanted to make their city a new gateway to New England, a bridge company was organized in 1871. Some work had begun in 1873; but after five years, financial problems had brought the project to a halt. A new firm, the Manhattan Bridge Company, took over in 1886 and contracted with the Union Bridge Company of New York City to design and construct the span. Thomas C. Clarke headed Union Bridge and was the firm's chief engineer for the bridge. Work was underway again before the end of the year.

To provide the clear spans needed for navigational clearances, the cantilever form was adopted for the main river crossing. It was one of the earliest major cantilever spans built in North America; and in both overall length and the length of its clear spans, the Poughkeepsie bridge substantially exceeded anything that preceded it. And unlike the several earlier cantilevers, which were built either partially or entirely with wrought iron, the Poughkeepsie bridge was built entirely of steel.

Another drawing from the same issue of *Scientific American* shows how the suspended spans were erected by cantilevering them outward from the completed anchor spans. Travelers on top of the bridge lifted each member into place from the river below.—Author's collection.

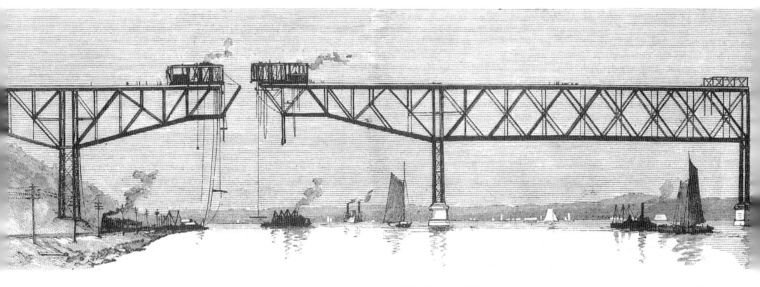

Across the Waters 37

This view of Poughkeepsie Bridge from Highland, New York, dates to sometime after the bridge was strengthened during 1906-07. At that time an additional line of trusses was added in the center of the span and new columns were added to the supporting towers.—Library of Congress.

The river crossing of the double-track structure spanned 3093 feet 9 inches center-to-center of its anchor piers and was supported by six intermediate piers. Piers on each shore supported cantilevers that projected 159 feet beyond the face of the supporting latticed steel towers, with 191-foot shore arms. Four piers in the river supported two enormous connecting span trusses, each 80 feet deep and 30 feet wide and spanning 525 feet between pier centerlines, with 159-foot cantilever arms at each end. Three suspended trusses between cantilever arms each spanned 212 feet. This provided two spans in the river with a full 500 feet between pier faces and a center span with 521 feet clear between piers. With the rail deck at an elevation of about 212 feet above mean high water level, the bridge afforded a clearance of 168 feet under the central suspended span and 130 feet under the main trusses, the minimum required by the bridge company's charter. Long steel approach viaducts at each end gave the bridge an overall length of 6667 feet.

Building the bridge's foundations in the deep, tidal waters of the Hudson represented the most difficult task for its builders. To reach a solid sand and gravel bed, the piers were taken down through 60 feet of water, mud and clay to depths of as much as 135 feet below high water. Instead of using either open cofferdams or compressed air caissons, the builders made a pioneering use of what was called the "open dredging" process. This was done by sinking an enormous open "crib" or caisson by dredging material from within it with "clamshell" or "orange peel" buckets without attempting to dewater the caisson.

Each of these cribs was made up of a grillage of foot-square timbers, typically 60 feet by 100 feet at the bottom and as much as 115 feet high. This was tapered at the bottom to form cutting edges. The inside of the crib was made up of a series of weighting pockets, closed at the bottom, and dredging shafts. Gravel was loaded into the weighting pockets to drive the crib downward through the river bottom as the mud and clay were removed through the dredging shafts. Once the crib had reached its final position on the river bottom, concrete was placed into the dredging shafts with clamshell buckets to form the underwater portion of each pier. These were brought to an elevation 20 feet below water level. Masonry piers of cut limestone were then built to an elevation of 30 feet above high water, and the steel supporting towers were then erected on top of them.

As they were being sunk into position on the bottom, the heavy cribs were held in place by anchors, each made up of a timber grillage filled with about 10 tons of stone. This worked well most of the time, but the builders ran into problems in placing the crib for one of the mid-river piers. The 5000 ton crib was being towed into position by three steamers in the spring of 1887 when a towing cable snapped after only one of the anchors had been placed. The heavy spring river flow was soon pulling the crib, three steamers, and the anchor downstream; and the entire assemblage had traveled 3 miles before it was finally brought under control. A second and third attempt were required before the crib was safely in place.

Erection of the steel superstructure was accomplished through a combination of scaffolding and cantilevered construction. The shore arms at each end of the structure were erected from timber scaffolding built on the ground. The two long connecting spans over the river were erected from timber scaffolding supported by piling driven into the river bottom. The scaffolding was built up to the level of the bottom of the trusses to provide a platform for a huge timber traveler which was used to assemble the great trusses. Once the side arms and main connecting truss spans had been assembled, the cantilever arms and suspended spans were erected from large travelers running on the top of the trusses by cantilevering them outward from the supporting towers.

One of the big L-1-a class 2-10-2 Santa Fe types that regularly worked the New Haven's Maybrook Line was in charge of a westbound freight on the bridge in this view from the late 1930s.—Charles A. Schrade, J. W. Swanberg Collection.

Landmarks on the Iron Road

While the Poughkeepsie Bridge no longer carries trains, busy rail lines cross under the span on both banks of the Hudson. A Conrail freight southbound on the West Shore line passed under the bridge at Highland on June 5, 1997.—William D. Middleton.

Thomas Curtis Clarke (1827–1901) was born at Newton, Massachusetts. He was schooled at home and at the Boston Latin School before entering Harvard at the age of 17. After graduating in 1848, Clarke took up civil engineering instead of studying law as he had originally intended. Over the next 18 years he worked in railroad location and construction with several railroads, as well as in architectural work, building construction and survey work.

Clarke's first major assignment in bridge engineering came in 1866, when he was engaged to build a crossing of the Mississippi River at Quincy, Illinois, for the Burlington. The project involved difficult foundation work and was the first iron bridge over the Mississippi, and Clarke's success helped to launch a distinguished career in bridge and foundation work. With Charles Kellogg, Clarke then formed the Kellogg, Clarke & Company bridge-building firm at Philadelphia. This firm lasted only two years, and in 1870 Clarke became the senior partner in Clarke, Reeves & Company, which soon became one of the leading bridge design and construction firms of the time. Clarke left this firm in 1883 to join in the formation of the Union Bridge Company at New York. In addition to the Poughkeepsie Bridge, Clarke's most notable Union Bridge projects included a crossing of the Hawkesbury River in Australia. Clarke's foundation experience proved particularly valuable for these projects, for both offered unusually difficult foundation conditions.

Clarke retired from Union Bridge in 1887 and spent the remainder of his life as a consulting engineer for bridge and other engineering projects. He served as president of the American Society of Civil Engineers during 1896–97 and was awarded both the Telford Medal and Premium of the British Institution of Civil Engineers for his 1878 paper on long-span iron railroad bridges. He died at New York.

The bridge was completed before the end of 1888; and on New Year's Day, 1889, the first scheduled passenger train crossed the Hudson on the new structure. The bridge and its connecting lines were soon acquired by the New Haven, and became an important new route between New England and the west.

The bridge proved remarkably adaptable for a structure of its time. The New Haven strengthened it during 1906–07 as heavier cars and locomotives came into use, inserting an additional line of trusses in the center of the span and adding new columns in the supporting towers. During 1917–18 the bridge was modified to accommodate double-headed operation of still heavier locomotives. This was done by installing a new floor system, further reinforcing of the main span and approaches, and by changing the double track on the bridge to a gantlet arrangement to avoid the eccentric loads on the structure that the larger locomotives would have created.

The bridge route remained an important link between New England and the west until well after World War II, but its importance began to diminish after the Penn Central merger in 1968, when most New England traffic began moving over the former Boston & Albany. Traffic over the bridge was down to one train a day when damages from a May 8, 1974, fire shut the bridge down. Repairs were never made, and the structure has stood there derelict ever since. In recent years, a volunteer, non-profit group, the Poughkeepsie-Highland Railroad Bridge Company, Inc., has launched an effort to establish a pedestrian and bicycle pathway on the bridge, which was listed on the National Register of Historic Places in 1979.

GETTING THERE

Poughkeepsie Bridge looms over downtown Poughkeepsie. Good views of the bridge are available from downtown Poughkeepsie, from the west bank of the Hudson at Highland, and from the parallel Mid-Hudson Bridge to the south. When a planned Walkway Over the Hudson project is complete, a walkway on the bridge will be accessible from the Highland end.

The Pennsylvania's Stone Arch Bridge (1902) at Rockville, Pennsylvania

As the Pennsylvania Railroad pushed westward from Harrisburg in the late 1840s, the Susquehanna River was its first major obstacle. At the Rockville site selected for the crossing, about 5 miles above Harrisburg, the river was shallow, with a rock bed, making foundation construction relatively easy. But the width of the stream necessitated a bridge almost three-quarters of a mile long

In this classic view of the completed Rockville Bridge, taken downstream from the west bank of the Susquehanna, photographer William Rau recorded all 48 arches and 3820 feet of the world's longest stone arch bridge.—Smithsonian Institution.

Some construction work for the interchange of the Pennsylvania's main line with the Northern Central line on the west bank of the Susquehanna was still in progress when this photograph was made at the west end of the bridge. The westbound freight train leaving the bridge will follow the main line up the river to its junction with the Juniata.—Smithsonian Institution.

that was the most costly structure on the new railroad. The first crossing at Rockville was a single track timber bridge. This lasted for almost 30 years before the demands of growing traffic required its replacement with a double-track iron span in 1877.

In 1900 the Pennsylvania began a massive improvement program that was needed to handle a steadily growing flood of traffic. Expansion of the main line to four tracks would be completed over the entire distance between New York and Pittsburgh. There would be a major new freight yard and other improvements for freight traffic in the Harrisburg area. To make it all work, a new bridge was again needed at Rockville.

Masonry bridge construction had fallen out of favor early in the era of railroad construction. But in the late nineteenth century the Pennsylvania turned again to masonry construction. Seeking structures of greater strength and durability, the railroad had begun to replace a number of its main line wood or iron bridges with stone arch structures, and this would be the choice for the third crossing of the Susquehanna at Rockville.

Designed by H. R. Righter under the direction of the railroad's chief engineer, William H. Brown, the new bridge was placed downstream from the old bridge, at right angles to the flow of the current. The bridge was 3820 feet long at the coping level and was made up of 48 arch spans of 70 feet each. The arches had a radius of 40

A high level view of Rockville Bridge from the hills above the east bank end of the Susquehanna shows an eastbound Conrail freight train leaving the bridge on August 21, 1978.—J. W. Swanberg.

feet and a rise of 20 feet above the springing line, which was placed at the level of the record flood of June 2, 1889, about 19 feet above the normal water level. The base of rail was 46 feet above normal water level, and the structure was 52 feet wide over the coping to provide room for four tracks. The bridge was wider at each end, to accommodate the beginning of curves to reach the railroad's lines parallel to the east and west shores of the river.

The structure was divided into six sections, divided by the abutments at each end of the bridge and five abutment piers. Every eighth pier was a 19-foot-wide abutment pier, while the intermediate piers were each 8 feet wide. The heavier abutment piers provided sufficient mass to withstand the horizontal thrust of the arches, allowing the bridge to be built in sections of eight arches at a time and limiting the potential extent of any failure in the event of flood damage to one or more arch spans.

Piers, arches and spandrel walls were built of white limestone quarried at a number of locations in the bituminous coal counties of western Pennsylvania. A number of new quarries were opened, and hundreds of men were employed to produce the 100,000 cubic yards of stone for the work. The arch rings were built with 42-inch thick limestone blocks, while the piers and spandrel walls were built with stones that varied from 18 to 24 inches. Piers, the spandrel walls above the piers, and the tops of the arches were filled in with Portland cement concrete. The concrete tops of the arches were covered with asphalt to provide a watertight surface, and the structure then was filled with cinders to the level of the track ballast.

The first stone was laid on May 1, 1900. Two contracting firms were engaged in the work, one working from each end. Cofferdams for each pier were sunk to the level of the rock bed of the river, usually no more than 3 feet below water level. These were then pumped out and a concrete foundation placed to the proper height to begin laying the stone piers. Once the piers were brought up to the springing line, timber truss centers were installed and the stone arch rings laid. Materials were delivered to the work by means of narrow gauge trains running on a temporary trestle alongside the bridge. Boom derricks mounted on barges were employed for handling materials for the piers and for placing the arch centers, while derrick travelers running on trestle work supported by the arch centers or the completed arches handled the arch and spandrel materials.

While construction was expected to take two years, unusually low water conditions helped to accelerate the work, and the structure was ready for service before the end of 1901. The related track changes, however, took longer to complete; and it was not until March 30, 1902, that the first train crossed over the bridge. The completed structure had cost the Pennsylvania somewhere around $1 million to build.

The Pennsylvania Railroad got the strong, enduring structure that it wanted; the massive stone arch bridge at Rockville has carried the trains of the Pennsylvania, and its Penn Central and Conrail successors, without interruption ever since. Even as Penn Central bridges above and below Harrisburg were wrecked in the record Susquehanna River floods that followed Hurricane Agnes in June 1972, the sturdy bridge stood firm.

At the time of its completion, the Rockville bridge

Eastbound train 40, Amtrak's successor to the Pennsylvania Railroad's flagship *Broadway Limited,* headed out across Rockville Bridge on April 29, 1984.—William D. Middleton.

was the longest railroad stone arch bridge anywhere in the world, and it still is today. The bridge was listed on the National Register of Historic Places in 1975 and was designated a National Historic Civil Engineering Landmark by the American Society of Civil Engineers in 1979.

GETTING THERE

Rockville Bridge is about 5 miles north of Harrisburg and just south of Marysville. The bridge can be seen from River Road on the east bank, while a back road between the river and the railroad leads south from Marysville to the west end of the span. Good general views are available from the U.S. highways that parallel the river on both sides.

Southern Pacific's Lucin Cut-Off (1904) across Great Salt Lake in Utah

The Southern Pacific's original route west of Ogden, Utah, left it with some difficult operating conditions. Blocked by the Great Salt Lake, the line diverged to the northwest to skirt the northern tip of the lake and then curved back to the southwest over a route of severe curves and steep grades to cross the summit of the Promontory range and a spur of the Hogup Mountains. By 1900, with Overland Route traffic growing steadily, the operating problems finally became too great. SP decided to build a new line right across the lake.

The cut-off planned by SP chief engineer William Hood (see page 92) ran almost due west from Ogden, crossing an arm of the lake to Promontory Point and then the main body of the lake on 32 miles of fill and trestle to reach the western shore at Strongs Knob, finally rejoining the original line at Lucin. The new line would be 102.5 miles long, cutting almost 44 miles in length, almost 11 full circles of curvature, and 1515 feet of vertical grade from those of the original line. The ruling grade on the cut-off would be only 0.5 percent, compared to 1.7 percent on the old line.

SP assembled massive stockpiles of materials and an enormous force of men and equipment for the work. Trainloads of timber and piling were shipped in. Eight big steam shovels worked the gravel pits and quarries opened up around the lake to supply fill material, while 25 huge pile drivers were shipped to Ogden in sections and then hauled out to the lake and assembled. SP gathered a fleet of 80 locomotives and close to a thousand cars for the work and even built a 127-foot sternwheeler, the *Promontory,* to serve as a supply ship and tender. Construction began early in 1902, and by summer more than 3000 men were at work. A station was established at each mile-end, where a boarding house for the workers was set up on piling. More than 330,000 gallons of fresh water had to be hauled in over a hundred miles of desert every day to supply the work crews. Work continued through the night under powerful electric lamps.

Two pile drivers started out from each mile-end station, working away from each other, back to back. On a 12-mile section of the line across the deepest part of the lake, where the water depth averaged 30 feet, more than 38,000 timber piles up to 130 feet long were driven to build a permanent, rock-ballasted timber trestle that would be the longest railroad bridge ever built in the U.S. Elsewhere, in shallower water, temporary trestles were built to allow work trains to dump rock fill for a permanent embankment.

Through the early part of 1903 the builders fre-

During 1902 and 1903 more than 3000 men were at work on construction of the cut-off across Great Salt Lake. This is a view of work on the 12-mile trestle section.—Union Pacific.

quently completed more than a thousand feet of trestle every day. But soon they began to have trouble with the unusual material at the bottom of the lake. Generally, the mud at the lake bottom was covered by a heavy crust of salt, soda and gypsum that was expected to be strong enough to support an embankment. At some points the gypsum crust proved so tough that piles could not be driven through it, and they had to be cut through with steam jets. At others, piling disappeared from sight in the mud bottom after a single blow of the pile hammer.

The real problems, though, were encountered when the SP began to place fill for the embankment from the temporary trestling. The material spread out across the lake bottom in the buoyant heavy brine, and ton after ton of rock and gravel seemed to disappear without a trace. At one point the fill material spread as far as 300 feet or more on either side of the track. At some points the crust gave way under the weight of the fill; at others the mud beneath the crust seemed to give way. On at

Workers spiked timbers into place during construction of the cut-off trestle in 1902.—Southern Pacific.

Steam power still prevailed on Lucin Cut-Off when an SP photographer recorded this scene of a freight train picking up orders at Midlake on the long trestle across Great Salt Lake. This unusual station was discontinued in 1946 after SP installed centralized traffic control on the cut-off.—Southern Pacific, David P. Morgan Memorial Library.

Across the Waters 43

In a classic photograph from the early streamliner era, a three-unit, porthole-windowed Electro-Motive E2 diesel sped across the cut-off's long trestle with the westbound *City of San Francisco,* jointly operated between Chicago and the West Coast by the Chicago & North Western, Union Pacific, and SP.—Southern Pacific, David P. Morgan Memorial Library.

least two occasions the embankment sank below the water level under the weight of a work train. The only answer was more fill. A force of 2500 men labored for months hauling it to the site, and gradually the embankment was brought up to the desired elevation.

SP president Edward H. Harriman drove a gold spike at a formal dedication of the cut-off on Thanksgiving Day 1903, but it was not until the following March that the track was stable enough to begin diverting trains from the old line. Even then, the struggle with the lake was far from over. At the west side of the lake the embankment continued to settle on almost a daily basis, sometimes by more than 4 feet at a time, through the end of 1904. The only answer was to dump still more rock fill into the lake.

In 1904 *Scientific American Supplement* termed the completed cut-off "perhaps the most noteworthy engineering achievement ever attempted in bridge-and-fill

Construction details of the rock-ballasted cut-off trestle are shown in this isometric drawing prepared by Robert J. McNair for the Historic American Engineering Record in 1971.—Library of Congress (Neg. LC-USZA3-117).

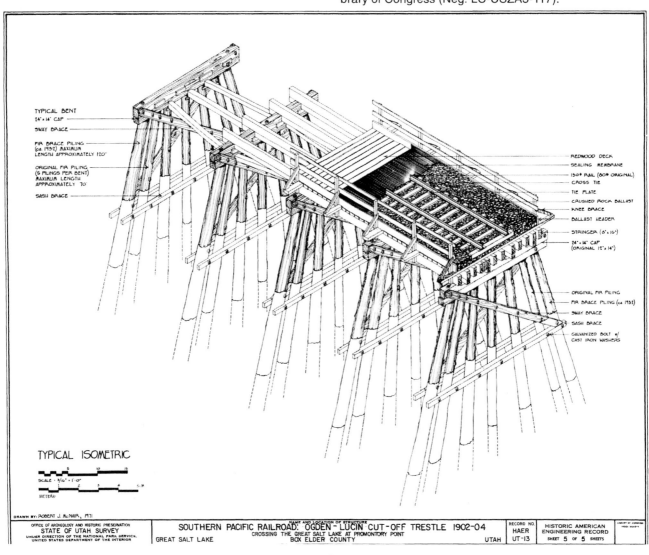

44   Landmarks on the Iron Road

work" and concluded that "the Southern Pacific, by building the Ogden-Lucin cut-off, has crowned the science of long distance trestling-and-fill embankment construction." The SP's battle with the Great Salt Lake has continued ever since, as the lake has risen and fallen on a cyclical basis. Construction of the cut-off coincided with the lowest water level ever recorded in the lake. By 1924 the water had risen by some 9 feet, and work trains shuttled thousands of carloads of heavy rock fill to the site to build up the line. In 1955 SP began a four-year effort to bypass the aging trestle with a 12.7-mile fill that required more than 31.5 million cubic feet of earth and rock.

By the mid-1980s the lake was on the rise again, reaching a level almost 16 feet above where it had been in 1903 and adding a west arm of the lake that extended the SP's water crossing by some 14 miles. For four years the railroad worked to repair and raise the embankment against the rising waters. Fierce storms in June 1986 washed out 11 miles of line, and the cut-off was closed for 77 days while more than 2 million cubic yards of rock and fill were brought in to rebuild it.

Despite the problems, Lucin Cut-Off proved a good investment for SP. From the very first the railroad reported substantial operating cost savings, while the direct, low-grade line has given the railroad the capacity to handle traffic growth that would have been impossible over the original line. The Lucin Cut-Off Trestle was placed on the National Register of Historic Places in 1972.

GETTING THERE

Located due west from Ogden, Lucin Cut-Off is largely inaccessible from public roads. Promontory Point, which is crossed by the cut-off, can be reached from a partially unimproved road that extends south from a point a few miles east of Promontory and about 25 miles northwest of Brigham City on State Route 83.

The Lackawanna's Tunkhannock Viaduct (1915) at Nicholson, Pennsylvania

Early in the century the Delaware, Lackawanna & Western began an extensive program of improvements designed to increase the capacity and operating efficiency of its main line between Hoboken, New Jersey and Buffalo. Chief among these works was the 40-mile Summit Cut-Off in northeastern Pennsylvania, where the railroad substantially reduced its grades and curvature by building an entirely new line across the rugged Appalachians that cut across hills, streams and valleys with massive cuts and fills, a tunnel, and two enormous viaducts.

Construction of the Lackawanna's great viaduct consumed an unprecedented volume of concrete. Two mixing plants churned it out at a rate of 40 cubic yards an hour, and a total of 167,000 cubic yards went into the structure. Here, workmen place concrete delivered by an overhead cableway into the base of one of the arches.—Nicholson Public Library.

After careful study of both construction and anticipated maintenance costs, the Lackawanna had chosen reinforced concrete as the most economical material for many of the structures required for its line relocation work, and the two viaducts on the cut-off were notable for their use of concrete on an unprecedented scale. Lackawanna chief engineer George J. Ray was credited with developing the overall plan for the cut-off, while Abraham B. Cohen, the railroad's concrete bridge engineer, designed the huge concrete structures.

The largest of these great works in reinforced concrete was a viaduct over Tunkhannock Creek at Nicholson. This massive double-track structure was 2375 feet long overall, stood 240 feet above the bed of the creek, and was 36 feet wide over the copings. Ten 180-foot arches and two 100-foot abutment arches, buried within the approach fill, made up the viaduct.

The arches were supported by piers that were carried as much as 95 feet below ground level to a foundation on bedrock. These massive piers were 36 feet 6 inches by 44 feet at the base, reducing to 28 feet by 34

This enormous 240-foot-high timber tower supported the center of the overhead cableway used for materials handling. At the time this construction photograph was taken on June 11, 1915, it was being used to place the steel arch centers for placement of the final arch at the center of the long viaduct.—Nicholson Public Library.

feet at the springing line of the arches. The main arches were semicircular, with an intrados radius of 90 feet and an extrados radius of 111 feet, tapering from a depth of 17 feet at their base to 8 feet at the crown. Each span was made up of two 14-foot-wide arch rings, separated by a space of 6 feet. These supported transverse walls of varying thickness, which in turn supported 13-foot 6-inch spandrel arches that carried the concrete floor system. The shorter abutment arch spans were of similar, if less massive, design.

While the viaduct was built of modern reinforced concrete, the engineers carefully designed the arches in the classical, semicircular Roman form. Each arch rib was scored to simulate the appearance of the voussoirs and key stone of a traditional stone arch, while similar scoring on the faces of the piers gave the appearance of coursed masonry.

Contractors began work in 1912, and the huge structure took more than three years to complete. Cofferdams for most of the pier foundations were sunk to bedrock without any unusual difficulty, but quicksand gave the builders problems at two of them.

A pocket of quicksand was encountered at the corner of one pier. This was overcome relatively easily by dividing the work into two sections and completing one end of the pier that was in solid ground, and then bracing against the completed concrete to hold back the soft material while the other end of the pier was completed.

At another pier, where a stratum of quicksand over the full area of the cofferdam was encountered about 75 feet below the surface and 20 feet above bedrock, the contractor had a much more difficult time. An attempt to drive sheet piling inside the cofferdam to divide the area into 24 sections, each of which could be completed separately, was unsuccessful. Finally, the builders turned to the use of a compressed air caisson to sink the pier the remaining distance to solid rock successfully.

Once the piers were complete, the main arch spans were placed through the use of self-supporting steel arch centers. Each set of centers, which supported a single arch rib, was made up of four three-hinged arch trusses which were supported on benches on the sides of the piers. After the centers had been used to complete one arch rib, they were struck, or lowered, and jacked laterally a distance of 20 feet into position to form the second rib. This complete, the centers were then dismantled and moved to the next span. Once the ribs were completed, wooden forms were used to place the lateral walls, spandrel arches, and deck. The smaller abutment arches at each end were built on wooden arch centers supported by wooden towers.

Arch centers and construction materials were handled by a combination of derricks and an overhead

Artist Joseph Pennell (1857–1926) was fascinated by the great man-made works of his time. "The mills and docks and canals and bridges of the present are more mighty, more pictorial and more practical than any similar works of the past; they are the true temples of the present," he wrote in "The Wonder of Work," published in *Scribner's Magazine* for December 1915. Inevitably, then, Pennell was drawn to the valley of Tunkhannock Creek, where he created this powerful etching, "The Lackawanna Viaduct," in 1919.—Library of Congress (Neg. LC-USZ62-104969).

Like some great Roman aqueduct, the completed viaduct loomed over the valley of Tunkhannock Creek and the little town of Nicholson, Pennsylvania.—George Arents Research Library at Syracuse University.

cableway more than 3000 feet long. The cableway was supported by timber towers 160 feet high at each end and a center tower 240 feet high, which was later raised to 280 feet as the work progressed. Concrete was hauled from mixing plants in cars pulled by dinky engines and was then moved in bottom dump buckets on the cableway to wherever it was required.

Day after day, the contractor's two mixing plants, each producing about 40 cubic yards an hour, churned out the concrete needed for the enormous structure. The volume of concrete placed reached as much as 14,000 cubic yards in a single month. The last concrete was finally placed in the viaduct on September 8, 1915. Tunkhannock Viaduct had required a total of 167,000 cubic yards of concrete, as well as some 1140 tons of reinforcing steel; and it was the largest concrete structure in the world.

Completion of the viaduct brought work on the Summit Cut-Off to a close, and the railroad celebrated its opening with a grand ceremony at the viaduct on November 6, 1915. The new line enabled the Lackawanna to cut 20 minutes from its through passenger schedules. Freight trains were able to save a full hour in getting over the division, and helper requirements were

Across the Waters

Soon after the viaduct's completion, a photographer recorded this interesting juxtaposition in transportation forms. In the foreground, a farmer's horse and buggy headed up the valley, while beyond a steam locomotive pulled a long northbound freight across the viaduct.—Nicholson Public Library.

substantially reduced. With traffic over the line up to record levels, the cut-off would prove to be a sound investment for the Lackawanna.

Still in use today by Canadian Pacific, Tunkhannock Viaduct remains the largest concrete railroad structure ever built. It was designated a National Historic Civil Engineering Landmark by the American Society of Civil Engineers in 1975 and listed on the National Register of Historic Places in 1977.

GETTING THERE

Tunkhannock Viaduct is located in the town of Nicholson, about 16 miles north of Scranton. Nicholson is on U.S. 11, or it can be reached via State Route 92 from Exit 64 on I-81.

The New York Connecting Railroad's Hell Gate Bridge (1917) at New York

A bridge across the East River at Hell Gate was the final element of the Pennsylvania Railroad's New York tunnel and terminal project begun in 1901. This linked the New Haven main line with both the new Pennsylvania Station on Manhattan and an improved freight route across New York Harbor from Bay Ridge in Brooklyn. Construction of the bridge and its connections was carried out by the New York Connecting Railroad, a subsidiary of the two railroads.

The bridge extended from Port Morris in the Bronx across Randalls and Wards islands and then across the East River to Queens on Long Island. Part of an important shipping route between New York Harbor and Long Island Sound, and thence the Atlantic, the East River channel at Hell Gate was 850 feet wide, and it had been dredged to a minimum depth of 26 feet.

Gustav Lindenthal, the chief engineer for the bridge,

developed comparative designs for continuous truss, cantilever, and suspension bridges before finally selecting an arch design with a clear span of about 1000 feet. This would assure the minimum required 135-foot clearance above mean high water and within the 700-foot width between bulkhead lines established by the War Department.

Lindenthal designed a two-hinged arch for the bridge, that is, an arch continuous between two end points which were hinged, or free to rotate. The arch itself was of the spandrel-braced type, widest at each end and curving gracefully up to the center of the span. Spanning between two massive masonry towers, this design was felt to be most expressive of the great rigidity expected in a railroad bridge.

The bridge was designed to carry a Cooper's E60 loading on its four tracks, and was carried by two arch trusses spaced 60 feet apart. The main arch span was 1017 feet between the faces of the masonry towers at track level and 977 feet 6 inches between the centers of the arch hinges at the abutments. The centerline of the parabolic bottom chord of the arch rose to a height at the center of 220 feet above the arch hinges. The depth of the arch trusses varied from 140 feet at the arch hinges to 40 feet at the center of the span, giving the structure a maximum height of 305 feet above mean high water. The bottom chords of the arches, their principal load-carrying members, were 6 feet 6 inches wide and varied from 10 feet 10 inches deep at the abutment bearings to 7 feet deep at the crown of the arch. Altogether, the bridge required 18,900 tons of structural steel.

The towers at each end of the arch were built of reinforced concrete faced with granite and stood 220 feet above the abutment foundations. They served both aesthetic and practical purposes. They helped to give the bridge a monumental appearance, and their massive size conveyed a feeling of the solidity that was needed to contain the thrust of the great arches.

Construction began in July 1912 for the massive foundations that would support the towers as well as the weight and thrust of the great arches. The work was relatively easy on the Long Island side, where the builders were able simply to rest the foundation on solid bedrock. Conditions were less satisfactory on the Wards Island side, where an array of 21 pneumatic caissons was sunk to bear on bedrock or clay at depths of as much as 127 feet below ground level and then filled with concrete.

Erection of the arches began in 1914. Because of the heavy East River traffic, the use of temporary scaffolding to support the arches during construction was impossible. Instead, Lindenthal erected the two arches outward from each tower as cantilevers. These cantilevered sections were held in place by temporary backstays, which in turn were anchored to huge counterweights on the ground. The arches were built, piece by piece, by traveling cranes which moved back and forth along the top chords of the completed arches. It took the builders only six months to "close" the arches, after which the backstays were removed and the arches began to carry their own weight. As the remaining work was completed on the bridge itself, contractors completed the long approach spans of steel and concrete that would carry trains up to the bridge deck, some 135 feet above the waters of the East River. By early 1917 the bridge was ready for service. There was a brief dedication ceremony on March 9, and a few weeks later the first regular train operated across Hell Gate.

At the time of its completion, Hell Gate Bridge ranked as the world's longest railroad arch bridge, and it still is. Now owned by Amtrak, Hell Gate remains a vital link in the passenger railroad's busy Northeast Corridor.

Erection of the steel arches for the Hell Gate Bridge was nearing completion in September 1915 as the two erection travelers placed the last members in the top section. Visible on either side are the temporary tie-backs that held the arch sections in place until the arch was closed and capable of supporting itself.—Pennsylvania State Archives, MG-286, Penn Central Railroad Collection.

Artist Joseph Pennell captured the drama of bridge construction in his 1915 etching, "The Bridge at Hell Gate."—Joseph Pennell (American, 1857–1926). The Bridge at Hell Gate, 1915. Etching, W670, 8 3/8" x 10 7/8". Gift of Persis D. Judd and Children, Picker Art Gallery, Colgate University, Hamilton, N.Y. Acc. No. 1977.26. Photo by Warren Wheeler.

GETTING THERE

Hell Gate Bridge can be viewed from Wards Island Park in the East River, which can reached from the Triborough Bridge or a foot bridge from Manhattan near 104th Street. On the Queens side, Astoria Park, reached from 21st Street, lies under the bridge. Good views are also available from the parallel suspension span of the Triborough Bridge.

—Smithsonian Institution (Neg. 90-8565).

Gustav Lindenthal (1850–1935) was born at Brunn, Austria (now Brno, Czech Republic). After studying at technical schools there he spent several years in engineering work on the Austrian and Swiss railways before emigrating to the United States in 1874.

Lindenthal soon began what was to be an exceptional career as a bridge designer and builder. By 1881 he had taken up his own engineering practice at Pittsburgh. In 1884 Lindenthal advanced a plan for a great suspension bridge over the Hudson River at New York, an idea he continued to pursue for the next 40 years. When the Hudson River project failed to materialize, Lindenthal went on to other major bridge projects, among them the Queensboro and Manhattan bridges over the East River, before the Pennsylvania commissioned him to design and build the Hell Gate Bridge. During this same period he also designed the enormous continuous truss bridge over the Ohio River at Sciotoville, Ohio.

When the Hudson was finally bridged at New York in 1931 by the George Washington Bridge, it was designed by Othmar H. Ammann, who had worked under Lindenthal on the Hell Gate and Sciotoville bridges. Lindenthal assisted Ammann as a design consultant, and the old man rode with Ammann to the dedication. He died at his Metuchen, New Jersey, home just four years later.

Northbound up the East River to Long Island Sound on its overnight run to Fall River, Massachusetts, the Fall River Line's elegant side wheeler *Priscilla* is seen passing under Hell Gate Bridge in this view from Queens. Built by the W & A. Fletcher Company of Hoboken in 1894, the *Priscilla* was still in service when the steamship line ended operations in 1937.—Library of Congress (Neg. LC-D-18-73418).

A New Haven class EF-3 electric locomotive headed the southbound Boston-Washington *Washingtonian* across Hell Gate Bridge in June 1949.—Collection of Amos G. Hewitt, Jr., J. W. Swanberg Collection.

The Canadian Government Railways' Quebec Bridge (1917) at Quebec

After seeking a crossing of the St. Lawrence for nearly half a century, success finally seemed at hand for the City of Quebec at the end of the nineteenth century. The Quebec Bridge Company was finally ready to begin construction of a span that would link the railway lines on both sides of the river.

The bridge site was 7 miles upstream from the Quebec Citadel, at a point where the river narrows to less than three-quarters of a mile in width, and flows between cliffs that stand more than 150 feet above high water. It was the best location there was on the lower St. Lawrence, but even so it was a difficult challenge to the skills of bridge engineers, who would have to deal with a maximum water depth of 190 feet and currents that flowed at close to 7 mph during ebb tides.

The bridge company appointed bridge engineer Theodore Cooper as its consulting engineer and contracted with the Phoenix Bridge Company of Phoenixville, Pennsylvania, to design and build a cantilever bridge with a main span of 1800 feet. This would rank as the longest cantilever span in the world. By the summer of 1907 the structure was well advanced when a collapse of the south end of the span plunged 75 workmen to their death in the river below. The failure was attributed to defective design and errors in judgment by the engineers.

A year later the Canadian government appointed a board of engineers to try again, chief among them American bridge designer Ralph Modjeski. Tenders were called based on a new design in 1910, with the bidders invited to submit alternate designs as well. Canadian, American, British, and German firms submitted proposals for a variety of cantilever designs, as well as steel arch and suspension alternatives.

A cantilever design proposed by the St. Lawrence Bridge Company of Montreal was adopted. This had anchor arms of 515 feet, with cantilever arms extending 580 feet beyond the main piers and supporting a 640-foot suspended center span, giving the bridge the same record-breaking 1800-foot clear span planned for the previous design. The trusses were spaced 88 feet on centers, allowing sufficient space for two tracks. The cantilever trusses reached a maximum depth of 310 feet over the piers, sloping to a depth of 70 feet at the ends of both the shore and cantilever arms.

This time, the engineers proceeded with extreme caution. The bridge was designed to support much heavier loadings than those used for the earlier bridge; train loads were based on Cooper's E60 loading. Extensive tests of materials were made, and design calculations were checked and rechecked. Even though the designers used a newly developed nickel alloy steel that could safely support stresses 40 percent greater than could the carbon steel used in the earlier design, the critical lower chord members of the anchor and cantilever arms were made several times larger and heavier than those of the failed bridge.

Some work was started at the end of 1909, and construction of the foundations began the next year. The south main pier was built with an enormous timber caisson 180 feet long, 55 feet wide, and 100 feet deep. This was built in a dry dock at Levis, towed to the site, and sunk into position. The north main pier was built

After first erecting the anchor arms on scaffolding, the builders of the Quebec Bridge erected the bridge's great 580-foot cantilever arms as self-supporting structures, building them piece by piece outward from the piers.—Smithsonian Institution.

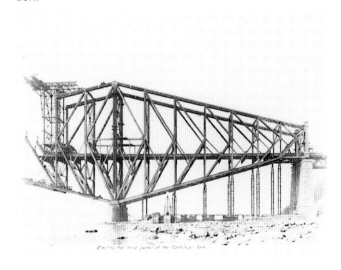

—*Smithsonian Institution.*

Ralph Modjeski (1861–1940) was born in Bochnia, Poland, and emigrated to the United States in 1876. After touring with his actress mother Helena for several years as her stage manager, Modjeski entered the Ecole des Ponts et Chausées at Paris in 1881. He graduated at the top of his class in civil engineering in 1885 and returned to the U.S. to begin a brilliant career in bridge engineering.

Over the next several years Modjeski worked with such well known bridge engineers as George S. Morison and Alfred Noble on major long-span crossings of the Missouri and Mississippi rivers. He established his own consulting engineering practice in 1893 and was soon engaged in major bridge projects in almost every part of the U.S.

Ralph Modjeski had become one of North America's preeminent long-span bridge engineers by the time he was appointed to the board of engineers for the Quebec Bridge, and he went on from that successful effort to design other notable bridges, among them such major suspension bridges as the Ben Franklin Bridge at Philadelphia and the Ambassador Bridge at Detroit. Modjeski's last and greatest work was his design for the great double-suspension and cantilever truss San Francisco–Oakland Bay Bridge, which opened in 1937, just three years before his death.

Massive and seemingly indestructible, the great Quebec Bridge has stood astride the St. Lawrence for more than 80 years.—*Canadian National.*

with two smaller caissons, each 60 feet by 80 feet. The caissons for both piers were sunk to a solid bearing material and filled with concrete to a level of about 5 feet below the extreme low water level. The main piers, which were 32 feet deep by 160 feet wide at the top, were built of cut stone facing and filled with concrete. Piers for the anchor arms, approach spans, and abutments were of similar construction.

Steel erection began in 1913. The anchor arms were erected by a traveler which moved outward from each shore to install temporary scaffolding and then to erect the trusses. The travelers then continued outward to erect the cantilever arms by cantilevering them outward from the piers. Work was nearing completion in 1916 when disaster struck again. As the 640-foot, 5000-ton

Made up of low-slung, tilting LRC equipment, VIA Rail's Montreal-Quebec train 22, the *Citadelle,* left the bridge on the north side of the river on its way into Ste. Foy and Quebec on September 21, 1993.—*William D. Middleton.*

Landmarks on the Iron Road

The Quebec Bridge's 640-foot, 5000-ton suspended center span was erected elsewhere, floated into position on barges, and then lifted into place between the cantilever arms by hydraulic jacks. This view shows the completion of the difficult task on September 17, 1917. A year earlier, an initial attempt ended in failure when the center span plunged into the river, killing 11 workmen.—Notman Photographic Archives, McCord Museum of Canadian History (Neg. 6081).

center span was being lifted into place between the cantilever arms from barges, a cast steel bearing failed, allowing the span to fall into the river. Eleven workers were killed this time.

A year later a new span was successfully lifted into place, and in October 1917 the first train finally crossed the bridge. The Prince of Wales helped celebrate its completion at formal opening ceremonies on August 22, 1919.

The bridge has stood firmly astride the St. Lawrence ever since, helping to link the Maritime Provinces and eastern Quebec with the rest of Canada. The record its builders set in 1917 still stands, for it remains the longest railroad cantilever bridge ever built. In 1987 the Quebec Bridge was designated an International Historic Civil Engineering Landmark by the American Society of Civil Engineers and the Canadian Society for Civil Engineering.

GETTING THERE

Quebec Bridge is 7 miles upstream from downtown Quebec, between Sainte-Foy and Charny. Provincial Route 132 crosses the bridge, and it can be viewed from Route 73 on the parallel Pierre-Laporte suspension bridge. An excellent view point is located at the Quebec Aquarium at the Sainte-Foy end of the bridge.

The Burlington's Ohio River Bridge (1917) at Metropolis, Illinois

While the Chicago, Burlington & Quincy was largely oriented to the west from its Chicago base, the railroad reached southward as well. Then under the control of the Great Northern's James J. Hill, the CB&Q began pushing south through Illinois to the Ohio River early in the century. Its goals were both to gain access to the coal fields of southern Illinois and to establish a gateway to the southeastern states and the Gulf Coast through Paducah, Kentucky. By 1910 the railroad had reached the Ohio River at Metropolis, Illinois, and plans were ready to continue across the river to a junction with the Nashville, Chattanooga & St. Louis at Paducah.

The Paducah & Illinois Railroad, which was jointly owned by CB&Q and NC&StL, was organized to build a new Ohio River bridge and the rail line between Metropolis and Paducah that were required to complete the connection. Design and construction of the mile-long bridge were under the direction of the Burlington's bridge engineer, Charles H. Cartlidge, while consulting engineer Ralph Modjeski supervised the design work and approved all plans and specifications.

The combination of navigational requirements for the river, and the Burlington's desire to build a structure that could handle the heaviest possible locomotive and car loadings, led to the design and construction of a formidable span. The railroad planned a double-track bridge and designed it for the live load of a train pulled by two Cooper's E90 locomotives, perhaps the heaviest loading standard ever used. The length of its main span would have to meet the War Department's requirement for a clear channel width of 700 feet.

The usual choice for a clear span of this length would have been a cantilever bridge. The Burlington's comparison of cantilever and simple truss span designs indicated that a cantilever could be built at substantially lower cost. By eliminating a need for falsework in the main channel of the river, a cantilever design would also have a significant advantage of safety in erection.

Despite these apparent advantages, however, the Burlington's Cartlidge strongly favored the use of a simple truss span for the main crossing. The principal reason for this was the greater rigidity that could be achieved with a simple span, but foundation conditions were also a factor. Foundations on rock were out of the question, since rock was at an average depth of 230 feet below the low water level. Instead, the piers would have to be founded on a thick stratum of white quartz sand that lay 75 feet below low water. This raised a possibility of differential settlement of the piers, which could introduce unanticipated stresses in a cantilever or continuous structure, while it would have little effect on a simple truss span. The simple truss span favored by Cartlidge was adopted, and, in order to meet the main channel clearance requirements, the bridge would include the longest simple truss span ever built for a railroad bridge.

The enormous Pratt trusses designed for this record span were 720 feet long. The two trusses rose from a height of 80 feet at the first panel point to a maximum of 110 feet at the center and were spaced 37 feet apart to provide room for two tracks. The span provided a 53-foot clearance for river traffic at high water level.

The balance of the river crossing on the Illinois side of the main span consisted of four similar through Pratt truss spans each spanning 551 feet and a single truss spanning 300 feet. A single 246-foot deck truss span was located at the Kentucky end of the main span. Long steel approach trestles at each end of the bridge completed the

The enormous pneumatic caisson used for construction of Pier No. 2 was built inside a floating dry-dock or pontoon on the Illinois bank of the Ohio. Still inside the pontoon, it will be floated into position in the river.—Smithsonian Institution (Neg. 98-2557).

The record 720-foot Pratt truss main span at the Kentucky end of the Metropolis Bridge was erected on temporary scaffolding in the river. The work was almost complete at the time of this photograph late in 1916, and the truss will soon be "swung" to carry its own weight, and the scaffolding removed.—Harold B. Hall Collection.

Only months after the record span was completed, a southbound Burlington train steamed across the Metropolis Bridge on March 13, 1918.—Smithsonian Institution (Neg. 98-2554).

5442-foot structure. Although designed for two tracks, the structure was completed with only one.

Some preliminary work was started in 1914. The substructure contractor built the seven river piers with enormous pneumatic caissons, one as large as 60 feet by 90 feet and almost 28 feet deep. Each caisson was built within a floating dry-dock or pontoon at a construction dock on the Illinois side and then floated into position, still inside the pontoon.

Once in position, the pontoon was submerged to release the caisson, which was then sunk to the required depth. Built of 12-inch-by-12-inch timbers, the caissons had a system of steel trusses that supported the roof of the working chamber. When the caisson was in position, a 19-foot concrete slab was placed over this roof to spread the load of the pier. The entire operation of sinking the caissons and building the concrete piers was conducted largely from work barges moored at each pier location.

During the construction of these deep caissons C. H. Cartlidge died suddenly of pneumonia in June 1916, which some attributed to his frequent descents into the pneumatic caissons to supervise the work. Ralph Modjeski thereafter took on the responsibility for supervision of construction as well.

Several of the piers were among the deepest ever built by the pneumatic caisson process. Water depths of over 110 feet were recorded for three of them, with one reaching a record-breaking depth of 113.2 feet below water level. Aside from the death of chief engineer Cartlidge, there were only two deaths attributed to the caisson work, a considerable improvement over most previous pneumatic caisson or tunneling projects of a similar nature.

The largest of the piers was 173 feet high over its base in the river bed and required 15,700 cubic yards of concrete. To provide adequate support in the sand and gravel river bed, the base of the pier was spread out over an area 110 feet long by 60 feet wide. The pier itself narrowed at the top to 65 feet long and 19 feet wide.

The contractors made steady progress, and by September 1916 both approach viaducts were complete, all

The disparity in size between the 720-foot main span and the four adjacent 551-foot spans is evident in this May 1997 view of the bridge from the Kentucky shore of the river.—William D. Middleton.

Three Union Pacific diesels are dwarfed by the scale of the enormous main span. The northbound unit coal train was returning empty from a nearby TVA generating plant in May 1997.—William D. Middleton.

but one of the piers were either complete or nearing completion, and one of the short trusses had been erected. By this time the superstructure contractor had begun building the timber falsework for erection of the long truss over the main river channel. This was completed by mid-October and a derrick car and cranes were then used to erect the huge truss. The work went smoothly and the enormous truss was safely completed only two months later. The first train crossed the completed structure on December 15, 1917, and the Burlington had its long-sought gateway to the south.

The Burlington's new line across Illinois to Kentucky prospered for a time, but with the changing railroad map of more recent years it no longer represents the major gateway its builders had anticipated. Today, only a few daily trains of successor Burlington Northern Santa Fe cross the Ohio on the record-breaking span.

But if its Paducah Gateway never quite lived up to the Burlington's expectations, the bridge at Metropolis was to become part of an important new north-south route for the Illinois Central. Initially, the IC was not interested in participating with the Burlington and NC&StL in building the new bridge and continued to use its own ferry crossing. But when the Ohio River froze solid in the bitter winter of 1917–18, the IC changed its mind. Unable to ferry cars across the river, the railroad asked to use the new bridge, and in 1923 became a joint owner of the P&I with the other two railroads.

IC became the principal user of the bridge in 1928, when the railroad completed its new Edgewood Cutoff. Operating via the Metropolis bridge between junctions with the IC main line at Edgewood, Illinois, and Fulton, Kentucky, the new line gave the railroad an alternate low-grade main line for fast freight service that was practically free of curves and 22 miles shorter than its original main line. Dominated by IC traffic ever since, the Metropolis bridge remains a key link in the movement of north-south freight traffic through mid-America, and its main channel crossing still stands as the longest simple truss railroad span ever built.

GETTING THERE

The north end of the bridge can be viewed from the river bank in downtown Metropolis, which is on U.S. 45, 3 miles west of Exit 37 on I-24. The record-breaking main span is at the Kentucky end of the bridge and can be reached from Paducah by following Route 1420, which parallels the river from the northwest side of town, west to a point just east of a P&I overpass, where a road leads north to a boat launching ramp with a good view of the bridge.

The Chesapeake & Ohio's Ohio River Bridge (1917) at Sciotoville, Ohio

Early in the century the C&O began expanding beyond its traditional eastern coal markets. The Midwest was becoming a major coal customer as well, and the railroad began seeking a route of its own into this new market. In 1911 the C&O gained control of Ohio's Hocking Valley Railroad and then incorporated a new subsidiary to connect it with the C&O's Cincinnati line in northern Kentucky. This was achieved with N&W trackage rights between Waverly and Columbus, Ohio, and by building a 30-mile line north across the Ohio River to Waverly.

The new crossing of the Ohio at Sciotoville would require another of the mammoth structures that were needed to bridge the broad waters of the river. Some unusual site conditions, together with the C&O's choice of one of the most accomplished bridge engineers of the time, created the opportunity for an innovative bridge of unprecedented scale. Gustav Lindenthal, who was then completing the steel arch span across the Hell Gate at New York, was chosen as consulting engineer for the Sciotoville bridge. It was to prove an inspired choice.

"It was often said of Mr. Lindenthal during his lifetime, and with truth, that he never built two bridges alike," commented the author of an American Society of Civil Engineers memoir following his death. "In this fact lies the best expression of his habit of looking on each bridge problem as new and unique, a problem whose proper solution could hardly be the same as that of any prior bridge problem.... A gift of creative originality was obviously essential for such a procedure, and this gift he possessed in high degree."

By February 20, 1917, erection of the bridge had passed the halfway point. The south span on the Kentucky side of the river was being erected by cantilevering outward from the center pier, while the north span had been erected on temporary scaffolding.—Smithsonian Institution.

Surely this characterized Lindenthal. For the Queensboro Bridge over the East River at New York he chose a double cantilever structure. His Hell Gate Bridge was an arch. The bridge across the Hudson that he promoted for 40 years would have been a suspension bridge. As with each of these, the Sciotoville bridge had its own special site conditions, and Lindenthal chose a design best suited to them.

Chief among the special problems posed by the site was one created by its location on a short curve in the river. During low water conditions, the navigation channel was on the inner or Kentucky side of the curve, but it shifted to the Ohio side during flood stage. Consequently, two long spans were needed for navigation clearances. The War Department required that each of these provide a 750-foot clear width. The site had an important advantage as well, and that was the presence of solid rock at only about 10 feet below the mean low water level at the center of the river.

This was a situation tailor-made for a continuous truss bridge. In a bridge made up of two or more simple truss spans, each of which rests freely on its end supports, each truss carries its load independently, gaining no support from the adjacent spans. But with a truss that is continuous over the intermediate supports, the load is distributed over the entire structure. This can produce substantial savings in material.

Despite the potential advantages, however, engineers had rarely used the continuous bridge form because of concerns over the unanticipated stresses that could be created in such a structure through a differential settlement of the supporting piers. The availability of a solid rock foundation at Sciotoville obviated these concerns, and Lindenthal took full advantage of this to design a continuous truss bridge that was a significant departure from typical American practice. He later estimated that the use of a continuous design had saved anywhere from 15 to 20 percent in the costs of materials and erection over what would have been required for a simple truss crossing. "Since the continuous truss is practically unknown in American bridge practice except for its use in swing bridges," commented *Engineering News*, "the radical character of the Sciotoville design is evident."

Lindenthal designed a truss that spanned 1550 feet center-to-center of the end piers and was divided into two equal spans of 775 feet. It would be the longest continuous truss bridge in the world. The trusses, which used the Warren system, were 129 feet 2 inches high at the center pier, where bending moments would be the greatest, reducing to a height of 103 feet 4 inches through the midpoint of each span and 77 feet 6 inches at the ends. To provide ample room for a double-track line, the trusses were spaced at 38 feet 9 inches, center-to-center. The design provided a minimum clearance of 40 feet above the river's high water mark and a low water clearance of more than 106 feet.

From a vantage point on the Ohio hillside at the Sciotoville end of the structure, a photographer recorded this splendid view of the bridge soon after its completion.—Smithsonian Institution.

The structure was designed for the live load of a Cooper's E60 train on each track. This heavy loading, together with the length of the span, necessitated some enormous truss members. The largest were the top chord members over the center pier, which weighed 114 tons and were 4 1/2 feet deep and 4 feet wide, with a cross section of almost 500 square inches of steel. The massive U-frame floor beams were more than 7 feet 6 inches deep. Instead of the pin-connected eyebar sections for truss members in tension then used by most bridge engineers, Lindenthal designed the entire structure with riveted connections to assure the greatest possible rigidity.

A flock of birds was startled into flight by the rumble of a northbound C&O freight on the massive bridge in 1977.—Robert A. Pike, C&O.

Long approach viaducts at each end of the bridge were made up of plate girder spans supported by concrete piers and linked to the main span at each side of the river by 152 1/2 foot deck truss spans.

Construction of the main pier at the center of the river was begun in October 1914, using a cofferdam 79 feet by 127 feet and 14 feet high. A 10-foot space between inner and outer walls was filled with sand and gravel to hold the cofferdam in place, and excavation was carried out from derrick boats. The pier on the Ohio shore was completed by open excavation, while the Kentucky shore pier was sunk by open dredging through wells. Concrete for both the main river and approach viaduct piers was mixed in plants on each shore and in a floating plant. The entire substructure for the bridge was completed by October 1915, and erection of the main span began the following spring.

Erection of the bridge employed unusual methods. On the Ohio side the continuous truss was erected on falsework supported by temporary steel bents resting on concrete pedestals on the river bottom. In order to maintain an open area for river traffic, however, the Kentucky side of the span was built by cantilevering the structure outward from the center pier. Work proceeded simultaneously on both sides, with the Ohio side sufficiently in advance to counterbalance the weight of the cantilevered Kentucky side. To avoid having to design the structure for cantilevering the entire distance on the Kentucky side, temporary steel bents were erected at two points close to the Kentucky bank of the river. Portions of the Ohio side of the bridge were erected by a

A pair of GE diesels headed out over the Kentucky approach viaduct toward the great continuous truss bridge with a northbound CSX coal train in April 1997.—William D. Middleton.

bridge derrick car operating over the falsework, while the trusses were completed by a steel gantry 160 feet high that stood astride the structure on tracks installed on the falsework. The Kentucky side of the span was erected with a creeper traveler which moved along the top chords of the completed structure.

The erection contractor began work in the spring of 1916, and C&O coal trains began rumbling across the completed structure on July 31, 1917. Now a key link between north and south for CSX Transportation, the Sciotoville bridge is still the longest railroad continuous truss bridge ever built.

GETTING THERE

Sciotoville is about 5 miles east of Portsmouth on the north bank of the Ohio. The north end of the bridge is accessible from U.S. 52 and local streets. On the Kentucky side the bridge is visible from Siloam, about 5 miles east of South Portsmouth on U.S. 23; and the south end of the bridge can be reached with a little perseverance over unmarked and unimproved local roads.

The Huey P. Long Bridge across the Mississippi River (1935) at New Orleans, Louisiana

By the time it reaches New Orleans, the Mississippi is a formidable stream. Vast quantities of water flow past New Orleans on the way to the Gulf; and where it sweeps by the city in a great crescent, the Mississippi is well over half a mile wide. Well into the twentieth century the river remained a barrier to east-west railroad traffic at New Orleans crossed only by ferries.

There were at least two unsuccessful attempts to build a bridge late in the nineteenth century. Another was launched in 1914 by the Public Belt Railroad Commission. For more than a decade the commission, its advisory engineers, and others debated various plans for a tunnel, a low-level bridge with lift spans, or a high-level bridge. River navigation interests opposed a low-level bridge, and the War Department refused to permit one. By 1925 the commission had adopted a plan for a high-level cantilever bridge and commissioned Ralph Modjeski to design it. The War Department approved the plan and some work was started. Little had been accomplished when financing problems that followed the onset of the depression in 1929 halted the project. Before work could resume, the War Department intervened with new requirements that necessitated still another design. There were still more financing delays in the early depression years, and it was not until the end of 1932 that work began at last.

As it was finally planned and built, the bridge was located several miles upstream from downtown New Orleans on a straight reach of the river. Aligned at right angles to the stream, which flows northeast at this point, the main crossing was on a nearly north-south axis, with the end of the structure on the east bank of the river actually west of that on the west bank.

To meet War Department requirements, the main river span was located close to the right bank of the river. The required 750-foot horizontal and 135-foot vertical navigation clearances were provided by designing the main section of the bridge as a cantilever. This had 529-foot anchor arms and a 790-foot central span that was made up of two 145-foot cantilever arms and a 500-foot suspended span. On the New Orleans side, the river

For more than half a century, until the great Huey P. Long Bridge was completed, the Southern Pacific's Sunset Route trains had to be ferried across the Mississippi River to reach New Orleans. This is an SP train ferry at the New Orleans landing.—Library of Congress.

Across the Waters

The 528-foot simple truss span on the New Orleans side of the bridge was erected by cantilevering it from the already-completed central section of the bridge. Work was nearing the midpoint of the span, where a temporary support was being constructed, when this view was taken on March 18, 1935.—Smithsonian Institution.

crossing was completed with a single 528-foot simple through truss span and one 330-foot and three 267-foot deck truss spans. The Warren trusses for the principal spans were spaced 33 feet apart, providing room for two tracks, and reached a maximum depth of 125 feet over the intermediate piers of the cantilever bridge. Roadways for highway traffic were cantilevered outward on brackets from either side of the trusses. The structure was designed for trains of Cooper's E90 loading on both tracks, plus highway loadings.

Tracks climbed to the high level crossing from the low alluvial plains on either side of the river on long approach viaducts. In order to connect with the railroad lines on either side of the river, the New Orleans approach curved through 107 degrees in the downstream direction, while the approach on the opposite bank swung through 122 degrees in the upstream direction. In order to maintain a 1.25 percent ruling grade for rail traffic the length of the approach viaducts totaled 19,471 feet, giving the crossing an overall length of 22,996 feet (almost 4.4 miles), making it the longest steel railroad bridge in the world.

The most innovative and difficult elements of the project lay below the surface of the Mississippi. The river was subject to floods that posed extreme hazards to any structure, while the sedimentary materials that underlay the river bed were ill-suited for the massive pier foundations that were needed. There was no bedrock to be found; borings to depths of 300 to 350 feet found nothing but sand, silt, gumbo, clay, shells, and vegetable matter. Some of the early studies for the bridge recommended that the main piers should be founded anywhere from 225 to 250 feet below Gulf Level. Finally, the noted soils engineer Dr. Karl Terzaghi concluded that caissons for the four main piers could be founded at a depth of 160 to 170 feet in a stratum of fine sand that would safely support the loading imposed by the bridge.

The engineers had planned that the difficult pier

foundations would be built using open dredge cellular caissons of unusual size. Contractors, however, were allowed to submit proposals for alternate construction methods. The successful contractor chose to use an innovative "sand island" method of sinking and controlling the caissons that the firm had developed to construct the piers for the Southern Pacific's crossing of Suisun Bay at Martinez, California. This involved the construction of a large island of sand contained within a steel shell at the site of each pier. The caisson was then constructed entirely above water level on the island. As each new section was completed, the caisson could be sunk by open dredging through the island and into the river bed until the cutting edge reached its final depth.

The method had a number of advantages. The use of the sand island eliminated all of the difficulties normally encountered in trying to hold a large floating caisson in place and made it possible to continue work without interruption during periods of high water. Construction of the caissons entirely above water permitted better control of concrete quality and allowed the use of caissons built entirely of reinforced concrete, instead of ones of wood and concrete as originally planned.

To avoid the danger of scouring action around the piers during floods, when current velocities were as high as 6 mph, enormous protection "mattresses" were placed over each of the pier sites before the caissons were started. These were made of woven willow about a foot thick, and were 250 feet wide and 450 feet long in the direction of the current flow. These mattresses were woven in sections on barges moored in the river, and were slid into the river as each section was completed. They were then sunk into place on the river bottom by piling a layer of stone on top of them.

The sand islands for the four main piers were constructed with rings of steel plate supported by a circle of piling driven around each site to form a shell. The shells were then filled with sand dredged from the river bed. Concrete was placed in wooden forms erected on the sand island to form the reinforced concrete caissons. Each caisson segment was 65 feet by 102 feet in plan, 20 feet high, and divided into 15 dredging wells. Once complete, the caisson sections were sunk by dredging sand from the dredging wells with clam shell buckets. As the top of each segment was sunk to just above the level of the sand island, another 20-foot lift segment was started on top of it.

Each of the main pier caissons was carried to a depth of about 170 feet below Mean Gulf Level. A timber cofferdam was placed on top of each caisson to build a large concrete block 30 feet high that supported the 150-foot-high concrete piers. The pier at the west end of the cantilever span and one pier under the deck truss spans at the New Orleans end of the bridge were also built with caissons, although these did not need to be taken to the same depth. The remaining piers under the deck truss spans at the New Orleans end of the structure were supported by timber piles.

The last pier was completed in November 1934, but steel erection had begun earlier in the year as other piers were completed. The superstructure contractor for the main river crossing began work at the New Orleans end late in March. These trusses were erected with guy derricks that were moved forward on skids, installing truss members ahead of them as they went. Temporary supports held the trusses in place until they were complete.

The cantilever portion of the bridge was built by what was called the "balanced cantilever" method, which began over each of the main piers and proceeded in both directions so that the two sections counterbalanced each other. The anchor arm on each side proceeded in advance of the cantilever arm and was partially supported by temporary falsework. The suspended section of the cantilever span was cantilevered outward from the two cantilever arms, while the simple truss span at the New Orleans end was similarly cantilevered from the adjacent anchor arm over the first part of its 528-foot span and then temporarily supported by falsework. The fourth deck truss span on the New Orleans side, the final section of the river crossing, was erected in October 1935. The approach spans had been completed in 1934.

Although some work continued until early in 1936,

This view shows the newly completed Huey P. Long Bridge from the New Orleans side.—Smithsonian Institution.

A Southern Pacific steam locomotive pulled an eastbound train up the approach to the bridge from the west bank of the river.—Southern Pacific, David P. Morgan Memorial Library.

the bridge was substantially complete; and it was opened to service on December 16, 1935. Soon after it opened, the bridge was named for Louisiana's populist governor and later U.S. senator, Huey P. Long, who had been assassinated in 1935. Used by both the Southern Pacific and Texas & Pacific railroads, and now Union Pacific, the Huey P. Long Bridge has been a key link in east-west rail traffic across the southern U.S. ever since.

GETTING THERE

The Huey P. Long Bridge is about 7 miles west and upstream from downtown New Orleans. U.S. 90 crosses the bridge. River Road passes under the New Orleans end of the bridge, while State Route 541 passes under the span on the opposite bank.

The Baltimore & Ohio's Arthur Kill Lift Bridge (1959) near Elizabeth, New Jersey

Railroads often found it necessary to cross navigable waters. Unless the topography allowed a high-level bridge the answer was usually some type of movable bridge. The earliest bridges of this type were usually swing spans, which rotated 90 degrees on a center pier to swing parallel to the channel, allowing water traffic to pass through. The bascule bridge, a counterbalanced span which rotated to a vertical position like a medieval drawbridge, later became a popular type.

The vertical lift bridge, however, was the most widely used movable bridge for railroad traffic, particularly where long clear spans were required. These employed a movable span that could be lifted vertically on towers on each side to clear the navigation opening. The weight of the lift span was counterbalanced by weights attached to each end through cables running over sheaves at the top of each tower. Lift spans reached some prodigious dimensions. The largest of them all was built for a Staten Island Rapid Transit crossing of the Arthur Kill near Elizabeth, New Jersey.

Originally isolated on Staten Island, the SIRT had been acquired by the B&O in 1885 to gain access to the island waterfront. It was then linked to lines in New Jersey by a swing bridge on Arthur Kill. By the late 1950s it was time to replace the old bridge. Arthur Kill was an important route between Newark Bay and the Atlantic, and water traffic through the draw span had reached an average of some 2400 vessel passages a month, among them tankers as large as 100,000 tons. The bridge had become a hazard to navigation, and the B&O also wanted to haul heavier loads across the bridge.

The new structure designed to replace the old swing span was a vertical lift bridge that would provide an unobstructed channel width of more than 500 feet and a 135-foot vertical clearance above high water. The lift span for the new bridge was a 2000-ton, Pratt truss structure spanning 558 feet center-to-center of bearings, 70 feet deep at the center, and 43 feet deep at each end, with the two trusses spaced 27 feet center-to-center. The two towers were made up of four cross-braced steel columns spaced 33 feet apart in the transverse direction and 32 feet longitudinally, and they towered 224 feet above the water. An operator's house was provided above track level on the tower at the Staten Island end of the bridge. A total of 13 plate girder approach spans gave the bridge an overall length of 1647 feet.

A view of the Arthur Kill bridge with the lift span in the raised position conveys a sense of the enormous size of the structure.—Herbert H. Harwood, Jr.

Each end of the lift span was supported by 40 wire ropes, each 2 1/4 inch in diameter, which passed across four 15-foot diameter cast steel sheaves at the top of each tower to counterweights that were made up of steel boxes filled with concrete. Each sheave weighed 23 tons and was carried on two roller bearings. An auxiliary counterweight system compensated for the unbalanced weight of the cables as the counterweights moved up or down. More than 99 percent of the 2000-ton weight of the lift span was balanced by the counterweights, making it possible to raise or lower the span 104 feet in only two minutes with an electrical drive system at the top of each tower. An electrical interlock system assured that the two ends of the bridge remained level throughout its travel. Air buffers were provided on all four corners of the lift span to prevent undue shock as the span was seated on the piers, while similar buffers were mounted at the top of each tower to provide a smooth stop in case of overtravel of the span beyond its stopping point.

Getting the bridge built presented some difficult challenges for its builders. The new structure was sited parallel to and about 60 feet north of the old swing span, which had to remain in service until the new bridge was complete. The two main piers under the lift span towers were to be built with 62-foot-by-68-foot timber cofferdams carried down to a rock bottom in about 40 feet of water.

This foundation work went about as planned at the Staten Island end of the structure, but things went seriously wrong at the pier under the tower on the New Jersey side. A 24-foot-thick mass of concrete placed in the cofferdam failed to develop the required strength, but proved almost impossible to remove. Twenty steel caissons, each 30 inches in diameter, were then drilled through the unsatisfactory concrete and into the rock at the bottom of the channel to establish a new foundation. Steel H piling was then placed through the caissons and concreted in place to support a new base slab of concrete 9 feet thick.

American Bridge Company, the superstructure contractor, had to work in tight quarters. The approach spans at each end of the lift span were completed first, and each of these was then used as the base for the erection of a temporary 125-foot tower supporting a derrick that was used to erect the permanent towers at each end. The huge lift span itself was fabricated at the contractor's Ambridge, Pennsylvania, plant, assembled elsewhere on Staten Island, and then floated into place

Across the Waters

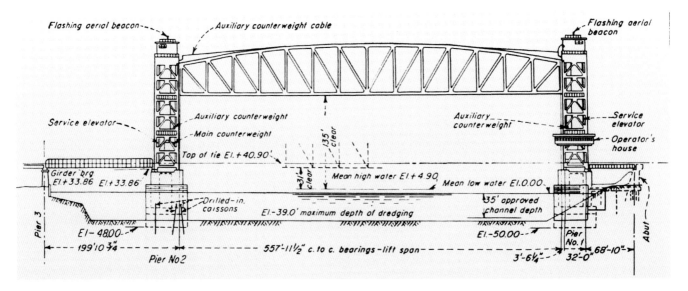

This drawing shows the general arrangement and principal features of the world's largest vertical lift bridge. The 2000-ton, 558-foot lift span was prefabricated elsewhere, then floated into place.—Reprinted from *Engineering News-Record*, June 11, 1959, © The McGraw-Hill Companies, Inc. All rights reserved.

on Arthur Kill and raised into position on the towers. Three large barges powered by four tugs were required to bring the huge truss to the site.

Foundation work for the structure had begun in 1957, the lift span was floated into place on May 31, 1959, and the bridge was completed by the end of the year. Water traffic was maintained throughout construction except for a six-day period when the lift span was floated in and raised into position.

For almost 30 years the bridge provided the only rail link between Staten Island and the mainland. The route continued to operate until 1986, when the Howland Hook Marine Terminal on Staten Island was closed, idling the huge Arthur Kill lift bridge.

More recently, however, the Port Authority of New York & New Jersey has reactivated the Howland Hook terminal and the railroad along the north shore of Staten Island to serve it. Early in 1998 contractors completed a rehabilitation and modernization of the bridge's electrical system and mechanical equipment, and the world's longest lift bridge was back in business.

GETTING THERE

The Arthur Kill bridge crosses from Elizabeth, New Jersey, to Arlington, Staten Island, New York, just a short distance south of the point where the Arthur Kill joins Newark Bay. There are good views of the bridge from the parallel Goethals Bridge used by I-278, just to the south.

The Burlington Northern's Latah Creek Bridge (1972) at Spokane, Washington

After the North American railroad system reached maturity early in the century, there were no longer very many opportunities for railroad bridge builders to challenge nature with daring new structures. The twentieth century brought such dramatic new bridge-building materials and construction technologies as prestressed concrete, welding, composite steel and concrete structures, and long-span precast segmental concrete girders. The suspension bridge form was pushed to record new span lengths, and such entirely new forms as the cable-stayed bridge were developed. But these have usually been applied to the structures of an expanding highway system or more recently to new rail transit systems, and only occasionally have they been seen on a railroad line relocation or bridge replacement project.

An exceptional opportunity to apply new bridge-building materials and technologies to railroad use came with the 1970 merger of the Burlington, Great Northern, Northern Pacific, and Spokane, Portland & Seattle to create the new Burlington Northern. The BN's first major construction project was a $16.2 million line improvement at Spokane, Washington, that would produce significant operating economies by linking the former NP main line through the city with the former GN and SP&S main lines to the west. Altogether, the project included 6.4 miles of new line and ten bridges, but its principal feature was a giant wye-shaped bridge,

Using what they called a "skyhorse," a large crane with a guyed boom, an erection crew lifted the first steel box girder section into the Latah Creek Viaduct in May 1972. The section was 80 feet long and weighed 82 tons.—Burlington Northern.

3950 feet long, that soared 210 feet above the Latah Creek canyon in West Spokane. The two curved branches of the wye were linked with the former GN line to Seattle to the north and the former SP&S line to Pasco to the south, while the main stem across the canyon joined the two lines to a former NP elevated line through the city.

Designed by Jack P. Shedd of consulting engineers Howard Needles Tammen & Bergendoff of Kansas City, the innovative structure employed a "composite" box girder design, in which steel was used for the bottom flanges and webs of the box section, while a reinforced concrete deck slab acted as the top flange of the box. The box girders were carried by tapered, rectangular reinforced concrete piers supported by spread footings of reinforced concrete founded on basalt or dense compacted gravel, except for those adjacent to the Interstate 90 highway, which were supported by drilled-in caissons.

The bridge included a total of 37 box girder spans, with 18 spans across the canyon, 10 in the north branch of the wye, and 9 in the south branch. Span lengths varied from 70 to 160 feet, and the box girders were anywhere from 7 to 12 feet deep, and 7 feet to 8 feet 6 inches wide. The girders were designed to function as continuous structures over four to six spans, supporting a Cooper's E80 loading.

The largest box girder sections were six 160-foot spans at the main canyon crossing. These were welded steel box sections built of 144-inch-deep web plates on each side and bottom flanges 8 feet 6 inches wide made up of steel plates and T sections. Internal stiffener plates and diagonals held the structure together in a rigid box shape. A composite structure, which employed the reinforced concrete deck slab as the top flange of the box girder, was achieved with welded stud shear connectors, which attached the deck slab firmly to the steel box. For both aesthetic and economic reasons, the steel box girder sections were fabricated with Bethlehem Steel Company's Mayari-R steel, a high-strength, corrosion-resistant material that weathered to a rich brown color and didn't require painting.

The girders for the bridge were welded together in sections up to 80 feet in length at the plant of a Kansas

Jurors for the American Institute of Steel Construction's 1973 annual competition for prize bridges called the Latah Creek Viaduct "light as a ribbon flowing across a deep valley." This is how the structure appears from the valley below.—HNTB Corporation.

City steel fabricator and shipped to the site by rail. These were then lifted into place from ground level using a large "skyhorse" crane with a guyed boom. Once the sections had been lifted into place, field connections were made with high-strength bolts, another modern development that has displaced riveting.

Work began on the first anniversary of the BN merger on March 2, 1971, and just 18 months later BN president Robert W. Downing and Spokane Mayor David H. Rogers joined to drive a golden spike marking completion of both the bridge and the new line. Dozens of BN trains every day were soon taking advantage of the new route, which cut almost 5 miles from the track distance between Chicago and Seattle and would save the railroad an estimated million dollars every year in operating costs.

In 1973 the innovative structure was named a prize bridge in the elevated highways or viaducts category in the annual competition conducted by the American Institute of Steel Construction to select the most beautiful steel bridges. "Simple yet majestic," said the competition jurors of the Latah Creek span, "this bridge appears light as a ribbon flowing across a deep valley. It is a breathtaking structure that provides a fine illustration of the effective use of modern materials and technology."

GETTING THERE

Latah Creek flows north into the Spokane River just west of downtown Spokane. The bridge crosses over the south end of High Bridge Park, which is reached by Sunset Boulevard. I90 is just south of the bridge and passes under the south end of the Y-shaped viaduct.

Heading west to the Pacific Coast, a Burlington Northern freight rumbled across the long box girder span.—Burlington Northern.

Four Southern Pacific Electro-Motive F3 units roared up the 2.2 percent north slope grade in California's Tehachapi Pass with eastbound train 2nd-806 in June 1954. Two more diesel helpers behind the caboose aided the long, hard pull up to the summit.—William D. Middleton.

2
Across Great Mountains

Railroad location, often as much an art as a science, was always a difficult task for railroad engineers, and it was particularly so as they sought the best possible routes across the high, rugged, and often uncharted mountain ranges of the North American continent. Locating an economical railroad route over relatively level terrain was one thing; achieving it successfully across a range of mountains was quite another.

In route location the engineer had to be concerned with both the immediate cost of construction and a line's prospective capacity and operating costs, two objectives that were often mutually exclusive. A line cheaply built to follow existing terrain might have steep grades and sharp curves that would both limit its capacity and create extremely high operating costs. Conversely, a line built with low grades and long radius curves for operating efficiency might require extremely costly cut and fill work, bridging and tunneling. Most North American railroads came down on the side of low construction costs, building their lines as quickly and cheaply as possible with the intent of later relocating or otherwise improving them to achieve better operating standards. Indeed, for many of them this process of upgrading lines to improve their operating characteristics has gone on for a century and more.

Engineers typically developed location surveys in three parts. First, a general reconnaissance was carried out to identify which of several prospective routes was the most promising. This was followed by a preliminary survey, a more accurate topographic survey of the general route selected. This usually established the maximum grade, determined the best line, and developed a detailed map and other data needed for final location. The last step was a location survey, a "final fitting of the line to the ground," as C. Frank Allen put it in his *Railroad Curves and Earthwork*. This laid out the plan for a line on the ground, ready for construction.

Few things could be more important to the success of a railroad project than the initial reconnaissance. A poorly chosen route that was unnecessarily costly either to build or to operate was a handicap that could not be easily corrected. A reconnaissance could be relatively easily and accurately done for a route being planned between well defined points, and for which accurate and complete topographical maps were available. But it was usually a much more difficult process for the pioneer engineers who built the early railroads across the vast western reaches of a North American continent that was often unsettled and largely unexplored.

A reconnaissance was typically done without survey instruments, using perhaps a compass for direction, an aneroid barometer to measure elevation, and some kind of odometer or pedometer to measure distance. One authority on railroad location, Willard Beahan, in his *Field Practice of Railway Location,* advocated "an experienced saddle-horse, whose speeds at his various gaits have been learned accurately by previous timing," as one way to do this rapidly.

Titled "Engineers in Camp," this drawing of a railroad survey party in the mountains, from Scribner's *The American Railway* (1892), portrayed an idyllic view of railroad location work.—Author's collection.

In the field for months on end, reconnaissance parties for the early railroads lived a rugged life, traveling on foot or on horseback and living in camps. Theodore Judah spent many months over a period of several years exploring the Sierra Nevadas of California before he finally found a route through Donner Pass for the Central Pacific. Eager to push their lines through as rapidly as possible, these reconnaissance parties often worked well into the winter. Searching for what would eventually form part of the route of the Canadian Pacific through the mountains of British Columbia, Walter Moberly kept his survey parties on the ground all through the winter of 1872–73. Fierce gales and temperatures 20 to 30 degrees below zero finally brought work to a halt. The snow was too deep for horses, and supplies were brought in to the surveyors at their winter camp by dogsled. John Stevens found the Great Northern's Marias Pass route through the Rockies on December 11, 1889, traveling alone on snowshoes in weather 40 degrees below zero.

The work could be hazardous, too. Early in 1853 the Congress authorized extensive surveys of possible routes for a Pacific railroad by the War Department. Later that year, while conducting one of the surveys, Army engineer Captain John W. Gunnison and six of his party were ambushed and killed by a band of Pahvant Indians at the Sevier River in Utah Territory.

One late nineteenth century essay on the building of railways by engineer Thomas Curtis Clarke for the June 1888 issue of *Scribner's Magazine,* however, painted an almost idyllic picture of railroad survey work:

> A surveying party will make better progress, be healthier and happier, if they live in their own home, even if that home be a travelling camp of a few tents. With a competent commissary the camp can be well supplied with provisions and be pitched near enough to the end of the day's work to save the tired men a long walk. When they get to camp and, after a wash in the nearest creek, find a smoking hot supper ready—even though it consist of fried pork and potatoes, corn-bread and black coffee—their troubles are all forgotten, and they feel a true satisfaction which the flesh-pots of Delmonico's can not give. One greater pleasure remains—to fill the old pipe, and recline by the camp-fire for a jolly smoke.

Walter L. Webb, in his *Railroad Construction,* took a more practical view, reminding a prospective engineer-in-charge of his responsibilities for the health and care of a survey party in the absence of nearby medical assistance. Webb's text was complete even to a recommended list of purgatives, emetics, and other medications that no well-outfitted survey party should be without.

Railroad location through mountainous terrain was a difficult task, requiring a skilled and experienced engineer to balance considerations of elevation, gradient, curvature, the amount of cut and fill required, and the balancing of cut and fill. John Bogart described it well in his essay on feats of railway engineering for the July 1888 issue of *Scribner's:*

> The location of the line of a railway through difficult country requires the trained judgment of an engineer of special experience, and the most difficult country is not by any means that which might at first be supposed. A line through a narrow pass almost locates itself. But the approach to a summit through rolling country is often a serious problem. The rate of grade must be kept as light as

The Deadwood Central Railroad Engineer Corps, photographed in 1888, was a rather dapper railroad survey party.—Library of Congress (Neg. LC-USZ6-7).

Titled "Taking a Level," this illustration of engineers at work locating a mountain railroad, from *Harper's New Monthly Magazine* for August 1874, gave a better sense of the arduous and often dangerous work of railroad location.—Author's collection.

possible, and must never exceed the prescribed maximum. The cuttings and the embankments must be as shallow as they can be made—the quantities of material taken from the excavations should be just about enough to make adjacent embankments. The curves must be few and of light radius—never exceeding an arranged limit. The line must always be kept as direct as these considerations will allow—so that the final location will give the shortest practicable economic distance from point to point.

Many a mile of railway over which we travel now at the highest speed has been a weary problem for the engineer of location, and he has often accomplished a really greater success by securing a line which seems to closely fit the country over which it runs without marking itself sharply upon nature's moulding, than if he had with apparent boldness cut deep into the hills and raised embankments and viaducts high over lowlands and valleys.

The most important of the location engineer's considerations were always elevation and gradient. The higher a train had to climb to cross a range of hills or mountains, the greater was the energy required to lift it to the summit. The steeper the gradient, the less tonnage a locomotive could pull. The engineer's principal goal was always to locate the lowest possible summit, combined with the best possible gradients on either side of it.

The Baltimore & Ohio built its pioneer line west to the Ohio at a maximum grade of 116 feet per mile, or 2.2 percent; and this was adopted by the Land Grant Acts of 1862 and later years for the Pacific Railroad and other land grant lines, specifying that "the grades and curves shall not exceed the grades and curves of the Baltimore & Ohio Railroad." Thus 2.2 percent became a maximum grade for the builders of the western railroads.

The principal earlier railroads of the eastern U.S. and Canada were usually able to do much better than this across the mountain ranges of the east. The Pennsylvania, for example, built across the Alleghenies with a maximum grade of 1.6 percent on the westbound grade out of Altoona to the summit. The rival New York Central avoided high summits and long grades altogether with its celebrated "Water Level Route," which followed the Hudson River and the shore of Lake Erie, but its route between New York and Chicago was more than 70 miles longer than the Pennsylvania's as a result.

The railroads of the west faced far greater challenges as they pushed across the Rockies to the Pacific. Most of the transcontinental lines had to climb to elevations of well over a mile to get across the Continental Divide, and many had grades of 2 percent or more. The highest of them all was the Denver & Rio Grande's crossing of Tennessee Pass in Colorado, which reached an elevation of 10,239 feet and had a maximum grade of 4.0 percent for eastbound traffic.

The lowest U.S. crossing of the Rockies was the Great Northern route through Marias Pass discovered by John Stevens, which was at an elevation of only 5213

Railroad Engineering: Grade, Curvature, and Cut and Fill

Grade: In North American practice, the rate of change in the elevation of a railroad line is typically expressed in feet per mile, or as a percent grade which represents the change of elevation in 100 feet as a percentage. A 1 percent grade, for example, is one that has a rise of 1 foot in each 100 feet. The English method states the grade in terms of the number of feet in which the grade rises a foot. Thus, a grade of 1 in 20 in the English system would be equivalent to a rise of 5 feet in 100, or a 5 percent grade.

The steeper a grade it has to climb, of course, the less tonnage a locomotive can pull, and railroad engineers attempt to limit grades to a maximum that will permit economical operation. Just how much a locomotive of given capacity can haul on a grade, compared to its capacity on level track, is a complex question that involves locomotive adhesion, desired operating speeds, and train resistance. To assist customers in selecting a locomotive, builder H. K. Porter Company published a number of tables of the approximate hauling capacities for its locomotives on various gradients. To cite just one example from one of these Porter tables, a locomotive with 80,000 pounds on driving wheels pulling a train of cars in good condition with an average train resistance of 8 pounds per ton could pull a train of approximately 1966 tons on level track. On a 1 percent grade the same locomotive could pull only about 532 tons. This potential capacity dropped to only 293 tons on a 2 percent grade, and to 196 tons on a 3 percent grade.

Main line railroads usually held their maximum grades to 2 percent or less except under the most severe conditions. The transcontinental railroads were generally able to keep their maximum grades to 2.2 percent or less, even in the most difficult sections of the Rockies. Perhaps the steepest grade on any current main line transcontinental route is the 2.4 percent maximum westbound grade on UP's former Rio Grande line across Soldier Summit in the Wasatch Mountains of Utah, southeast of Provo.

There were much steeper grades than these. Grades of 4 percent and more were commonplace on the narrow gauge railroads of Colorado and similar mountain lines elsewhere. Norfolk Southern still operates a former Southern Railway line over Saluda Mountain in North Carolina on a 4.7 percent grade that is the steepest main line grade anywhere in North America. Probably the steepest grade ever used in regular line haul service is one operated by short line Madison Railroad north out of the Ohio River valley from Madison, Indiana. Opened by the Madison & Indianapolis, a Pennsylvania Railroad predecessor, in 1841, the line's 5.9 percent grade up from the river was originally operated as a rack-and-pinion incline, but was converted to ordinary adhesion operation in 1868. Although it couldn't pull very much, an adhesion locomotive could climb grades of 10 percent or more under good track conditions. One extreme example was the B&O's temporary switchback line over the Kingwood Tunnel in 1851, which employed grades of 10 percent.

A **ruling grade** is usually established for each segment of line. This is not necessarily the maximum grade on the line, but rather the grade which limits the maximum tonnage a locomotive can haul over the line segment. A long grade, or one on which trains might have to stop and start, for example, can be more restrictive than a short, steeper grade than can be partially overcome by the momentum of a train. In order to maintain the lowest possible ruling grade over a line for reasons of overall operating efficiency, steeper intermediate **pusher** or **helper grades** are sometimes established. At these points helper locomotives are assigned to enable a road engine get the same full tonnage train over the steeper grade that it can handle unassisted over the balance of the line.

A **compensated grade** represents the combination of an actual grade with an equivalent additional grade representing the greater train resistance of a curved track, as discussed more fully under curvature below. In practice, engineers typically reduce the actual grade on curves in order to maintain a compensated grade equal to the established maximum grade for the line.

To avoid an abrupt change in slope as a train moves between line sections on different gradients, engineers usually introduce a parabolic **vertical curve** to provide a smooth transition between the two sections.

Curvature: Railroad curves are typically made up of arcs of a simple circle. In British or metric practice, curves are normally designated by their radius. But in North American practice, curvature is usually expressed in degrees, minutes and seconds of the central angle subtended by a straight line or chord of 100 feet, measured between two points on the curve, rather than by the radius of the curve. Thus, the shorter the radius and the sharper the curve, the greater the degree of curvature. A 1-degree curve, for example, has a radius of about 5730 feet; a 10-degree curve a radius of 573 feet; and a 60-degree curve a radius of only about 95 feet. A good rule of thumb is to divide 5730 by the degree of curvature to obtain the radius in feet.

A curve made up of segments of different degrees of

curvature or radii is called a **compound curve,** while a curve in which the direction of curvature changes from one direction to the other is called a **reverse curve.** Railroads normally use **spiral** or **easement** curves to provide a gradual transition and a smoother ride between straight or tangent line and curved track. A spiral is also used to gradually introduce **superelevation,** a raising of the outer rail in a curve to offset the lateral centrifugal force created as a train passes through a curve at speed.

Operation around curves, in addition to limiting train speeds, also increases the frictional resistance of trains, largely because of the binding of wheels mounted on parallel axles that are not exactly radial to the curve. This is partially offset by slightly widening the track gauge on curves, but a significant resistance remains. In a curved line on a grade, this additional curvature resistance is usually converted to an equivalent grade which, when added to the actual grade, establishes a compensated grade for the section. In practice, engineers usually compensate for grades on curves at a rate of about 0.04 percent grade for each degree of curve. Thus, a 5-degree curve on a 2 percent actual grade would add an equivalent grade of 0.2 percent to establish a compensated grade of 2.2 percent for the curved section.

Cut and Fill: The alignment for a line being built over irregular topography rarely coincides with the existing ground level. Instead, the line usually falls below or above, requiring some amount of cutting or filling to establish a roadbed at the required level. Excavation, as well as placing and compacting fill material, is expensive work, and still another concern of the location engineer is to plan an alignment that would keep such cut and fill work to a minimum.

However much or little excavation or filling are required to bring the roadbed to the desired level, the engineer tries, too, to plan a line so that the required quantities of cut and fill between adjacent sections of line are equal, or "balanced." This balance avoids the cost of either acquiring additional fill or disposing of excess excavated material during construction. Once a tentative horizontal and vertical alignment has been established for a line, the engineer can use topographic data and cross sections of the cut or fill sections to compute the volume of cut or fill for each segment of line. This is then developed into a "mass diagram," a graphic representation used to determine the balance of cut and fill, and average haul distances. This process might be repeated several times with different grade lines or alignments until the best possible balance of cut and fill and the most economical haul distances are achieved.

Early railroad building was done largely with hand labor, horses and mules. This was a group of workmen on the Union Pacific.—Library of Congress (Neg. LC-USZ62-35458).

In later years, mechanical equipment took on much of the heavy work of railroad building. This Bucyrus steam shovel was at work on the construction of the Western Pacific in 1906.—Library of Congress (Neg. LC-USZ62-34573).

One of the more spectacular examples of the art of railroad location in mountain terrain was the Union Pacific's celebrated Georgetown Loop, laid out by civil engineer Robert Stanton to achieve a workable grade in the steep canyon between Georgetown, Colorado, and the Silver Plume mining camp. This view of the Loop is from Scribner's *The American Railway* (1892).—Author's collection.

feet, with a maximum grade of 1.8 percent. This gave the GN a permanent operating advantage over its rivals for traffic to and from the Pacific Northwest. The earlier Northern Pacific crossed the Divide at 5566 feet in Montana's Mullan Pass and had an even higher pass to surmount west of Bozeman, where the line reached an elevation of 6328 feet in Homestake Pass. Maximum grades in the two passes were a 1.9 percent eastbound grade in Homestake Pass and a 2.2 percent westbound grade in Mullan Pass. The Milwaukee Road, the last to build west, had an equally difficult crossing, climbing to 6347 feet to cross the Divide in Pipestone Pass, where it faced maximum grades of 1.7 percent eastbound and 2.0 percent westbound.

Maintaining reasonable grades in steep, mountainous terrain was usually a difficult task. Sometimes the engineer could find natural terrain along a ridge or stream that provided a direct route at a satisfactory grade, but more often the slopes were too steep for this. When this was the case, the only way to achieve a satisfactory grade was to reduce the rate of change of elevation by lengthening the line.

This was usually accomplished by taking advantage of some natural feature of the topography to introduce a deviation that would lengthen the line. This might be achieved, for example, by running a line up one side of a lateral valley and back down the other, reversing direction with a sweeping curve. An arrangement like this created the Pennsylvania Railroad's celebrated "Horseshoe Curve" as the railroad climbed the east slope of the Alleghenies above Altoona, Pennsylvania.

Another way to accomplish this was the switchback. Instead of going straight up a slope, the additional length of line required to reduce the gradient was gained by

The arrival of the railroad builders brought intense, if transient, activity in their wake. This was the scene at the summit of Rogers Pass in 1886 as the Canadian Pacific was pushing its main line westward through the valley of the Illecillewaet. Businesses in the row of false-fronted shacks offered such necessities of life as laundry, wines, liquors, and cigars. The larger structure in the distance was the Pacific Hotel. In the middle distance at right, the workers are clustered around the pay car.— Ross & McDermid, Canadian Pacific Archives (Neg. NS.4742).

running the line backwards and forwards up a slope with a zigzag track arrangement. But switchbacks were always a time-wasting burden to the operating forces, since a train had to be stopped and reversed every time the line changed direction, and they were seldom employed except as a temporary expedient.

One early example of the use of switchbacks was in the construction of the B&O in 1851. To permit trains to continue west to the railhead even before the long Kingwood Tunnel in western Virginia (now West Virginia) was completed, the railroad's chief engineer, Benjamin H. Latrobe, had a series of steep switchbacks built right across the top of the tunnel alignment.

In 1886 the Northern Pacific used a similar arrangement to carry traffic across Stampede Pass in Washington's Cascade Mountains while a long tunnel under the pass was completed. Further north in the Cascades, James J. Hill's Great Northern operated across Stevens Pass from 1893 to 1900 on a series of switchbacks with 4 percent grades until the first Cascade Tunnel was completed.

Still another alternative was to develop some sort of loop or spiral, in which a line climbed through a complete circle, passing over itself with a tunnel or bridge. Among the best known North American examples are the Southern Pacific's Tehachapi Loop in California, or the Canadian Pacific's spiral tunnels in Kicking Horse Pass, British Columbia. Perhaps one of the most spectacular loops of all, however, was the famous Georgetown Loop built by the Union Pacific to reach the Silver Plume mining camp from Georgetown, Colorado. A direct line between the two points would have had to climb 600 feet in 1 $1/4$ miles, requiring a gradient of over 9 percent. By building a loop line that reversed itself four times and extended the distance to 4 miles, the railroad was able to reduce the maximum grade to only 3 percent.

The Western Railroad's Crossing of the Berkshire Hills (1841) in Massachusetts

Boston began early to build railroads to the west. The Boston & Worcester was organized in 1831, and only two years later the Western Railroad was formed to build the remaining 117 miles from Worcester to the New York state line.

The line from Worcester to Springfield was open by October 1839. The western end, where the railroad followed the Westfield River up into the rugged Berkshires Hills of western Massachusetts, was a much more difficult task. Between Springfield and the Berkshire summit the railroad had to build a line over which trains would be able to climb almost 1400 feet in some 39 miles. The Western's chief engineer for location and

—*Smithsonian Institution.*

George Washington Whistler (1800–1849) was one of the band of early railroad engineers educated at West Point. He was born at the Army post of Fort Wayne, in what is now Indiana, where his father was commandant. Following graduation from the Military Academy in 1819, Whistler taught for a time and then did survey work on the U.S.-Canadian border. His railroad engineering career began in 1828 on the B&O. Whistler first joined William McNeil and Jonathan Knight on a trip to study railroads in Europe, returning to build the B&O's first mile of track. He later worked with McNeil on several other new railroads in the Northeast and spent several years in lock and canal work at Lowell, Massachusetts.

Whistler returned to railroad work in 1836 with his appointment to the Western Railroad, where his extraordinary achievements brought him to the attention of the government of Russia, where Czar Nicholas I had decided to link Moscow and St. Petersburg with a 400-mile railroad. In 1842 Whistler accepted an invitation to plan and build the line. He spent six difficult years in Russia, and the great railroad was nearly complete when he was stricken with cholera and died at St. Petersburg, cutting short the career of one of the best of the early American railroad engineers.

After the death of his first wife, Whistler married Anna McNeil, the sister of his long-time professional colleague, in 1831. Among their five sons was the painter James Abbott McNeil Whistler, whose celebrated 1872 painting of his mother in old age, "Arrangement in Grey and Black No. 1," has become an icon of American art.

construction of the line was George Washington Whistler, an experienced railroad engineer, and he solved the problem remarkably well.

Some of the railroad's directors wanted to build the line cheaply, meandering up the hillsides to follow the existing contours, but Whistler insisted that a more direct line should be built, even if at greater expense, and that the cuts, embankments and bridges should be built wide enough for a future second track.

An 1840 report to the Massachusetts Legislature spoke of the extraordinary difficulties encountered. Ground water and unstable ground were frequent problems. At one point a long section of embankment settled from 40 to 45 feet below the natural surface of a meadow. The work of building two deep cuts west of Pittsfield was handicapped by constant freezing and thawing of materials in the winter; and, everywhere, heavy snows interfered with progress, sometimes requiring two to three weeks to clear.

The most difficult section of all was the 13-mile Summit Division east of Pittsfield, where the line followed the Pontoosuc River, the west branch of the Westfield, through a rocky, circuitous, and narrow defile to reach the Berkshire summit at Washington, 1440 feet above tidewater. Constructed at a cost of as much as $220,000 per mile, it was one of the most expensive stretches of railroad yet built.

Building the line to Whistler's high standards required deep rock excavation and the construction of 21 stone arch or timber bridges, some as high as 70 feet above the water, and with spans of up to 60 feet. Embankments as much as 70 feet high and heavy masonry river walls 70 feet high were required at some points. All told, this section of line required the excavation of some 440,000 cubic yards of earth and 183,000 cubic yards of rock. At the summit, the line passed through the ridge in a cut more than a half mile long and 52 feet deep at its deepest point, requiring the excavation of 57,000 cubic yards of rock of the hardest kind.

On one 2-mile section of line east of the summit, grades of more than 1.5 percent were required; eastbound traffic faced even heavier grades on the west slope, with one section near Pittsfield at almost 1.7 percent.

Two B&A steam engines helped the westbound Boston Section of the *20th Century Limited* up the Berkshire Hills grade at Washington Hill.—Clifford G. Schofield, David P. Morgan Memorial Library.

It was a bitterly cold morning on February 15, 1964, as six New York Central diesel units led a westbound Boston & Albany freight through an ice-encrusted Washington Cut at the top of George Washington Whistler's crossing of the Berkshire Hills.—Jim Shaughnessy.

Led by one of the railroad's ubiquitous Electro-Motive GP40-2 units, an eastbound Conrail freight rumbled across the West Branch of the Westfield River on Bridge No. 1 in June 1997 as the train descended the east slope of the Berkshire crossing toward Springfield and the Connecticut River. The venerable twin stone arch bridge, just above Chester, Massachusetts, is one of the line's original structures constructed by Alexander Birnie in 1842 under the direction of Major Whistler.—William D. Middleton.

The first grading work west of the Connecticut River had begun in March 1838. The line opened between Springfield and Chester in May 1841, and the first through trains over the completed line crossed the state line into New York in December 1841.

The Western Railroad was the longest yet constructed in America by a single corporation, and its mountain pass through the Berkshires represented an unprecedented achievement for the still-new field of railroad engineering. Writing of it some years later, Charles Francis Adams called it "the most considerable enterprise of its kind which had then been undertaken in America; and taking all the circumstances of time, novelty and financial disturbance into account, it may well be questioned whether anything equal to it has been accomplished since."

Whistler's crossing of the Berkshires later became part of the Boston & Albany. A major B&A line improvement in 1912 relocated portions of the line along the west branch of the Westfield but never very far from the original line. Today, the line is the principal rail freight route into New England, and Conrail diesels in a steady parade still haul tonnage across the Berkshires on the line laid down by George Washington Whistler more than a century and a half ago. The Middlefield-Beckett Stone Arch Railroad Bridge District, which includes some of Whistler's original structures among its seven surviving bridges, was named to the National Register of Historic Places in 1980.

GETTING THERE

Conrail's Berkshire crossing is southeast of Pittsfield, with the principal westbound grade located between Chester and Washington. Chester is reached by U.S. 20, and the beginning of the stone arch bridge section lies about 2 1/2 miles north on Middlefield Road. Whistler's famous summit cut at Washington can be reached by traveling east on Summit Hill Road from the Washington town hall on State Route 8.

The Pennsylvania's Crossing of the Allegheny Mountains (1854) west of Altoona, Pennsylvania

In 1846 the Pennsylvania Railroad was chartered by the Pennsylvania Legislature to build an all-rail line across the state from Harrisburg to Pittsburgh and the Ohio River. The task of locating and building the railroad fell to J. Edgar Thomson, an experienced railroad engineer, who was appointed chief engineer in April 1847.

There were earlier surveys for an all-rail route over the Alleghenies completed by Charles L. Schlatter in 1842. Schlatter had studied three general alignments and had recommended a middle route between Harrisburg and Pittsburgh. This would follow the Susquehanna northward and then the Juniata River to the west as far as Lewistown, crossing the summit of the Alleghenies west of Hollidaysburg at an elevation of 2200 feet. This, he said, would be the cheapest to build, the shortest, at 229.5 miles, and could be built with a maximum grade of 0.85 percent.

Schlatter's route would have begun to climb with grades at his 0.85 percent maximum grade west of Lewistown in order to achieve an easy crossing of the Allegheny summit. Thomson made his own surveys and

This early view shows a westbound Pennsylvania passenger train on Horseshoe Curve about 1870. Double-tracking of the entire line over the Allegheny summit was completed in 1855. During 1898–1900 third and fourth tracks were added on the busy line through the curve.—Hagley Museum and Library.

In a classic 1940 scene of Horseshoe Curve activity from the great years of the Pennsylvania Railroad, Class I-1s 2-10-0 Decapod No. 4517 storms through the curve on the outer track with a train of empty coal hoppers for the western Pennsylvania coalfields. —H. W. Pontin, Herbert H. Harwood, Jr. Collection.

chose a different route that followed the Juniata and Little Juniata rivers for most of the distance, establishing a 134-mile low grade route from Harrisburg to the foot of the mountains with no grade steeper than 0.4 percent. Then from a point at the foot of the mountains, where Altoona was later established, Thomson's line turned west to make a head-on assault on the Alleghenies with grades of as much as 1.8 percent. Based upon his operating experience, Thomson believed that the combination of a long, low grade line and a short, steep climb over the summit would prove much more economical to operate than Schlatter's location with its much longer segments at 0.85 percent.

Thomson's route over the summit climbed out of Altoona and up the valley of Burgoon Run on a 1.8 percent grade, and it then turned at Kittanning Point and wound around Burgoon Run to form the Pennsylvania's famed Horseshoe Curve. The 3-mile horseshoe crossed over the Run on a huge fill at the 9-degree curve itself, which extended through 220 degrees, more than half a full circle. The line climbed up to the curve on a grade of 1.75 percent, dropping to 1.45 percent in the great

from The American Railway (1892). —Author's collection.

John Edgar Thomson (1808–1874) was born in Delaware County, Pennsylvania. He was a largely self-taught engineer educated by his father, a surveyor who had worked on early canal and railroad projects. Thomson became a member of the state's engineer corps at the age of 19 and was soon engaged in railroad surveys.

Returning from a trip to study European practice in 1832, Thomson was appointed chief engineer of the Georgia Railroad. Fifteen years with the Georgia company gave Thomson a solid base of engineering and railroad operations experience that would suit him well for the demanding post as the Pennsylvania's chief engineer. In addition to his engineering work for the railroad, Thomson was actively involved in the efforts to finance the company, which undoubtedly had much to do with his election as president in 1852.

Thomson led the Pennsylvania during a 22-year period that saw the company well started on its development into the greatest of all American railroads. He completed the main line to Pittsburgh and then leased the line that pushed the system west to Chicago in 1869. Thomson brought the state's connecting roads east of Harrisburg into the system, and in 1871 he extended the Pennsylvania to New York with the lease of lines in New Jersey. All told, under Thomson's administration the Pennsylvania grew more than six-fold to a system of more than 1500 miles of road. Still active as president, he died at his home in Philadelphia a few months after his 66th birthday.

The Pennsylvania's conquest of the Alleghenies was completed with multiple tunnels under the summit. Here, a Class M-1 4-8-2 Mountain type freight engine leads an eastbound merchandise train into the west portal of the New Portage Tunnel at Gallitzin, Pennsylvania.—W. J. Pontin, David P. Morgan Memorial Library.

The line had reached the foot of the Alleghenies by 1850. Construction across the summit began the following year with work on the double-track Gallitzin tunnel, which turned out to be the most difficult part of the entire crossing. Four shafts were sunk to the level of the tunnel, and drilling was carried out from six faces. The tunnel was drilled through an unstable combination of coal, fireclay, and shale, which swelled, cracked, and fell when exposed to air. All drilling was by hand, and progress on each heading averaged only 15 feet a week. The tunnel workers encountered large volumes of water, and steam pumps had to be installed to remove as much as 175 gallons a minute.

Thomson became president of the Pennsylvania early in 1852, and completion of the main line to Pittsburgh fell to others. The railroad's Western Division, which extended from Big Viaduct, east of Johnstown, to Pittsburgh, was completed by the end of 1852. Only the opening of the Mountain Division across the summit was needed to complete the entire route between Harrisburg and Pittsburgh. All that was left was the difficult tunnel at Gallitzin, and this became the most urgent task for Herman Haupt, who was appointed chief engineer in 1853.

The first locomotive passed through the tunnel before the end of 1853, but rock falls delayed its opening. Haupt hurried the tunnel to completion, shoring up the roof with timbers where necessary, and the tunnel—and the Mountain Division—were finally opened to traffic on February 15, 1854. Traffic over the new route grew rapidly. Even before the full line was complete, the installation of a second track was begun; by 1855 the entire line over the Allegheny summit had been double-tracked.

Still more improvements came in the Pennsylvania's great decade of expansion at the beginning of the cen-

semicircular curve to compensate for the additional resistance of the curvature; and then it continued through the top half of the horseshoe on a 1.73 percent grade. From one calk, or heel, of the 3-mile horseshoe to the other, the rails climbed through a distance of 122 feet.

Above the curve, the line continued climbing, first to the south and then westward above Sugar Run. At Gallitzin, the railroad drilled the 3570-foot Allegheny Tunnel to cross the summit at an elevation of 2200 feet. From Altoona to the summit, Thomson's spectacular line climbed through 896 feet in a distance of just 9.8 miles. West of the summit, Thomson was able to establish a descending grade no greater than 1 percent for the line as it dropped down through Cresson to follow the Conemaugh River west toward Johnstown and Pittsburgh.

With some help from the State of Pennsylvania, Conrail completed tunnel enlargement and other work that enabled the railroad to begin moving double-stack container trains over the Alleghenies. One of the first double-stack trains rounded the Horseshoe Curve in 1995.—Conrail.

The Pennsylvania and Sylvania Electric teamed up to produce this unprecedented photo-flash picture to celebrate the centennial of Horseshoe Curve on October 20, 1954. The simultaneous flash of 6,000 flash bulbs illuminated the railroad's westbound New York-Chicago *Trail Blazer* coach streamliner in the foreground, while two freight trains moved around the curve.—Sylvania Electric–Pennsylvania Railroad, David P. Morgan Memorial Library.

tury. A new tunnel was completed under the summit parallel to the original Gallitzin tunnel in 1904, and a new double-track line was established east of the summit over the rebuilt New Portage Railroad. The Pennsylvania's route across the Alleghenies had become the busiest mountain line in the world.

In later years there was talk of a new, low grade tunnel and electrification, but nothing came of any of these plans. Until diesels took over in the 1950s, Horseshoe Curve and the Allegheny crossing remained one of the greatest shows in all of steam railroading.

Today's activity over J. Edgar Thomson's Allegheny crossing is but little diminished from the halcyon days of the Pennsylvania. The number of tracks around the curve has been reduced to three, but a steady parade of freight traffic still moves around the great Horseshoe Curve and across the summit. In 1995 Conrail, with some help from the State of Pennsylvania, completed major tunnel enlargement and other clearance improvement projects all across the state that brought double-stack container trains over the Alleghenies for the first time. Horseshoe Curve was placed on the National Register of Historic Places in 1966.

GETTING THERE

Altoona is located on U.S. 220 and U.S. 22, 34 miles north of the Bedford Exit on the Pennsylvania Turnpike. The Horseshoe Curve National Historic Landmark Visitors Center is on Kittanning Point Road (State Route 4008), west of Altoona. The summit tunnels are at Gallitzin, 8 miles west of Altoona on U.S. 22.

The Central Pacific's Donner Pass Crossing of the Sierra Nevada (1867) in California

The building of the transcontinental railroad was one of the great achievements of nineteenth-century American railroad engineers. No part of that extraordinary effort presented a more difficult challenge to its builders than did the Central Pacific's 105-mile route through Donner Pass to cross the Sierra Nevada (literally, "snow-covered mountains") of California.

There was no easy way through the Sierra Nevada's formidable granite peaks, but the CP's brilliant chief engineer, Theodore D. Judah, found the best that there was. In 1861 Judah joined with Collis P. Huntington, Mark Hopkins, Leland Stanford, and Charles Crocker to incorporate the Central Pacific; and in August of that year he completed the surveys that established its route across the Sierras.

Judah had been making surveys for a route over the mountains since 1856, initially for a wagon road but always with a railroad in mind. Routes through both Madeline and Hennes passes to the north were considered, but in 1860 a Dutch Flat druggist named Daniel W. "Doc" Strong suggested another that would prove the best of all. Instead of following a river valley, Strong's route followed an unbroken ridge between the south fork of the Yuba and Bear rivers and the north fork of the American River through Dutch Flat to Donner Pass. Shorter than any other route, the line Judah chose would

Across Great Mountains 81

—Union Pacific.

Theodore Dehone Judah (1828–1863) was born in Bridgeport, Connecticut. Soon afterward his family moved to Troy, New York, where he received most of his education, studying for a time at Rensselaer Polytechnic Institute. Judah soon began a distinguished career in railroad engineering, working for several eastern lines. In 1854 he became chief engineer for the Sacramento Valley Railroad and moved to California to take up the task of locating and building the new line. The first railroad on the Pacific Coast, the line reached Folsom early in 1856, but even before then Judah had become a tireless advocate for the great Pacific Railroad that would link California with the East.

Judah spent much of his time over the next several years in exploration and survey work in search of the best route across the Sierra Nevada. His fervent support for the Pacific Railroad took him frequently to Washington to press the case for federal support. He ultimately found his principal support from the "Big Four," Sacramento merchants Huntington, Stanford, Hopkins, and Crocker. Judah joined with them to form the Central Pacific and then returned again to Washington, where his efforts finally met with success with the passage of a Pacific railroad bill in July 1862. Not long after construction began early in 1863, Judah returned east once again, this time to raise money to avoid being forced out of the CP by his four partners. He contracted yellow fever during the passage across the Isthmus of Panama and died at New York early in November, leaving to others the task of completing the Pacific Railroad of which he had dreamed for so long.

Locating and building the Pacific Railroad through the rugged western mountains was hazardous and difficult in the extreme. This drawing, from *Harper's Weekly* for May 30, 1868, shows a Central Pacific survey party at work in Humboldt Pass in western Nevada.—Author's collection.

Bloomer Cut, west of Auburn, California, 800 feet long and 63 feet deep, was the deepest cut of all on the Central Pacific's line through Donner Pass. This stereopticon photograph by Alfred A. Hart shows the cemented boulders that were encountered by the construction crews.—Library of Congress (Neg. LC-USZ62-56967).

This sweeping trestle at Secrettown, 1100 feet long and 50 to 90 feet high, was one of the largest structures on the line across the Sierra Nevada range.—Library of Congress (Neg. LC-USZ62-56979).

The fierce winter of 1867–68 convinced the Central Pacific that some sort of protection would be needed for its trains, and ultimately some 40 miles of snowsheds were built. This drawing of the interior of a typical timber snowshed is from the May 30, 1868, *Harper's Weekly*.—Author's collection.

Photographer Alfred A. Hart recorded the hard rock interior of the Central Pacific's 1659-foot summit tunnel before the track was laid. The bore was drilled from both ends and a center shaft at an average rate of about 2 feet a day.—Alfred A. Hart, Library of Congress (Neg. LC-USZ62-56960).

An artist for the May 1872 issue of *Harper's New Monthly Magazine* illustrated the CP's new snowsheds in the depth of winter.—Author's collection.

Across Great Mountains

Another drawing from the May 1872 *Harper's New Monthly Magazine* depicts a train rounding one of the precipitous curves at Cape Horn. —Author's collection.

extend 81 miles from the base of the mountains east of Sacramento to the summit, climbing through nearly 7000 feet. But so well located was the line that the maximum grade could be held to just under 2 percent. Minor changes would be made during construction, but Judah's alignment was essentially what the CP followed, and no one has ever found a better route across the Sierra Nevada.

Judah died less than a year after work began in January 1863, and Samuel S. Montague, who had worked with him on the surveys, took over as chief engineer. Charles Crocker headed the great construction force that pushed the CP across the mountains.

The line was built largely by manual labor, with pick and shovel, horse-drawn carts, and black powder. Labor was in short supply, and in 1865 Crocker turned to Chinese laborers to meet the immense needs of the project. By the time the work was complete some 20,000 of them had been brought over from southern China. At the peak of construction in 1867, Crocker had more than 13,000 men and 1000 teams at work in the Sierra Nevada.

The work was hard going. Some of the embankments reached heights of 70 feet, and many cuts had to be blasted through solid rock and cemented boulders. Bloomer Cut, west of Auburn, was 800 feet long and 63 feet deep, while others were as long as 1200 feet and 40 to 50 feet deep. High timber trestles carried the line across deep ravines. The trestle at appropriately named Deep Gulch was 100 feet high and 500 feet long, while a great curved trestle at Secrettown was 1000 feet long and 50 to 90 feet high.

At Cape Horn, the rocky walls of a cliff rising abruptly above the American River were cut away to create a shelf for the roadbed 2500 feet above the river. To do it, workers were lowered from the top in baskets to pick and drill holes for black powder charges in the rock face. The workers were hauled up, the charges set off by fuses, and the process repeated.

Fifteen tunnels, 11 of them within a 20-mile section over the summit between Cisco and Lake Ridge, were laboriously blasted through the granite mountains at a rate of about 2 feet per day. One of the most difficult was the 1659-foot bore across the summit at an elevation of 7017 feet. It was drilled from both ends and from a shaft sunk at the center. At one point, Crocker had as many as 8000 men, working in three shifts, at work on the summit tunnels.

The fierce winters of the Sierra Nevada made the work even more difficult, and the winter of 1866-67 was the worst ever. There were some 44 snow storms, one of them a two-week gale which left 10 feet of snow on the ground. The total snowfall that winter was some 40 feet. Snow tunnels were cut so that work could continue on the summit tunnels. Snowslides swept whole buildings and dozens of men down the mountainsides, and one large timber trestle at Cisco was lost to an avalanche. The experience convinced the CP that some sort of

Stereopticon photographer Hart set up his heavy camera on the roof of a locomotive to record this unusual view of Lost Camp Spur cut, 80 miles east of Sacramento.—Alfred A. Hart, Library of Congress (Neg. LC-USZ62-51737).

protection against the snow was needed to operate the railroad successfully during the winter, and the builders were soon at work constructing some 40 miles of snowsheds.

The Sierra Nevada crossing was finally complete in 1867, and CP rails had reached the Nevada line by the end of the year. Crocker's construction crews then raced onward across Nevada and Utah to the meeting with the Union Pacific at Promontory Point on May 10, 1869, which marked the completion of the transcontinental railroad.

From that time until this, Overland Route trains have followed Judah's line through Donner Pass to cross the Sierra Nevada. The CP soon became part of the Southern Pacific system, which later began a program of improvements for the line over Donner Pass. Line relocations eased some of the grades, and by 1925 the entire line had been double-tracked.

In more recent years SP had removed some of that double track as traffic levels declined. But merger with Union Pacific in 1996 reversed those declining fortunes, and UP crews were soon at work upgrading track and enlarging tunnels to permit double-stack container train operation across the historic route. Today, as the trains climb through Donner Pass, they continue to follow the line laid down by Theodore Judah more than 130 years ago.

The Central Pacific Railroad was designated a National Historic Civil Engineering Landmark by the American Society of Civil Engineers in 1968.

A new era for Donner Pass is reflected by this view of a westbound Union Pacific coal train descending the west slope of the pass at Emigrant Gap in August 1998. With heavier traffic resulting from the merger of Southern Pacfic into UP in 1996 and new run-through agreements with Burlington Northern & Santa Fe, restoration of the second track removed by SP in the early 1990s was likely.—Stan Kistler.

GETTING THERE

The Donner Pass line is accessible from I-80, which generally parallels the railroad all the way from Sacramento to Reno. The most difficult sections of the Sierra Nevada crossing are largely on the western slope, which begins just east of Roseville. The summit is at Norden, just west of Donner Lake.

The Mexican Railway's Maltrata Incline (1872) on the Line between Vera Cruz and Mexico City

Mexico's first important railroad linked Vera Cruz, on the Gulf of Mexico, with Mexico City. Such a line had been proposed as early as 1833, but in the political turmoil of nineteenth-century Mexico it would be another 40 years before the trains were running. Surveys for a route between the two cities were finally begun in 1858 by Captain Andrew Talcott, yet another of the West Point graduates who were so prominent in early North American railroad engineering.

By early 1859 Talcott had completed a survey for a route between the two cities via Orizaba, and construction began soon after. Work was interrupted several times, and it was not until late 1872 that railheads from east and west for what was now the British-financed Ferrocarrill Mexicano (Mexican Railway) met near Maltrata. Through service over the 264-mile route was inaugurated on January 1, 1873.

By far the most difficult section of the Mexicano to build was the mountain division between Paso del Macho and Esperanza in the Sierra Madre Oriental, where the line climbed a total of some 6483 feet in only 64 miles to reach the central plateau of Mexico from the coast. In locating the line through the rugged area between Orizaba and Boca del Monte, Captain Talcott had determined that the reduced construction and maintenance costs for a shorter, but steeply graded, line would provide the most economical alignment, despite the higher operating costs, compared to a longer line with more favorable grades.

One of the Mexicano's Fairlie locomotives and a train of four-wheel construction flat cars negotiated the spindly Wimer Viaduct, above Orizaba in the Maltrata Mountains.—A. Briquet Photograph, Library of Congress.

The Ferrocarril Mexicano's Metlac Ravine Viaduct, near Fortin de las Flores on the railroad's climb up through the Sierra Madre Oriental, stood 93 feet high, and simultaneously negotiated a sharp curve and a 3 percent grade.—A. Briquet Photograph, Library of Congress (Neg. LC-USZ62-116981).

Close to a century later, the great Metlac Viaduct was still there, although heavier plate girders and stone piers had replaced the spindly original structure. A single General Electric box cab electric and two Electro-Motive F units headed westbound Vera Cruz–Mexico City train 52 across the viaduct on March 15, 1961.—Jim Shaughnessy.

In its major reconstruction and line relocation of the 1980s, the Mexican National Railroads (FNM) built a new Metlac Viaduct near Fortin which, at 1410 feet long and 430 feet high, is the highest railroad bridge in Mexico. Three General Electric U36C units headed an eastbound freight across the structure on February 14, 1991.—J. W. Swanberg.

The resulting alignment provided ruling grades that ranged from 1.75 to 2.7 percent compensated in the segment between Paso del Macho and Orizaba, where the ascent began in earnest. Over the next 15 miles the line climbed first through a 3.5 percent maximum grade and then went to ruling grades of 4.1 and 4.5 percent compensated. In a final 10 miles between Maltrata and Boca del Monte known as the Maltrata Incline the line reached ruling grades of 4.7 percent compensated, with some maximum grades as steep as 5.25 percent.

With the snow-capped cone of the 18,700-foot Orizaba Peak looming over it to the north, it was a spectacularly scenic railroad. Cut into the sides of the steep mountain slopes, almost the entire line was located on continuously reversing curves, many of them ranging from 12 degrees to as sharp as 16.5 degrees. Between Orizaba and Esperanza there were seven tunnels. Above Orizaba Wimer's Viaduct curved above a deep ravine on spindly metal towers. Near Fortin de las Flores the great Metlac Viaduct, a nine-span structure that stood 93 feet high, simultaneously negotiated both a curve and a 3 percent grade.

An article in the June 1874 *Harper's New Monthly Magazine* described the wondrous journey over the newly completed railway:

> High along the side of this exceedingly steep hill creeps the railroad, making some of the most surprising feats of engineering as it winds and leaps across this chasm. It becomes almost circular as it twists and turns.... The cars begin to climb up the Cumbres; four thousand feet they accomplish in less than thirty miles. It is holding on by the eyelids.... The road is the finest bit of engineering on this, if not on any continent.

For half a century the Mexicano attacked Talcott's prodigious grades with a fleet of British-built Fairlie locomotives, a unique articulated double-end design that was ideally suited to the steep grades and sharp curves of the Maltrata Incline. But even two of the Mexicano's largest 155-ton Fairlies of 1911 took 4 hours to move a 360-ton freight up the 4.7 percent grade, and in 1928 a fleet of big GE-Alco box cab electrics began moving tonnage up the Incline. Electric traction continued to dominate Mexico's big hill until the 1960s, when diesels began to take over under Mexican National Railways (FNM) ownership.

West of Paso del Macho FNM drilled a 689-foot tunnel through Mount Atoyac and linked it with this unique Pensil Viaduct-Tunnel. A double-stack container train negotiated the structure.—Mexican National Railways.

The former Mexicano offers the shortest route between the port at Vera Cruz and Mexico City, and FNM began a program of improvements during the 1980s to expand capacity and permit the operation of high-cube double-stack container trains. On the mountain division the line is being almost entirely rebuilt to provide a second main track and to establish new standards of a 2.5 percent maximum grade and 6-degree maximum curvature.

Between Los Reyes, at the western end of the division, and Ciudad Mendoza, FNM turned away from the short distance, steep grade philosophy set by Captain Talcott more than a century before to build a new 47-mile second main line via the Acultzingo Valley that is 18 miles longer than the original line but reduces the maximum grade and curvature to the new standards. To do it, FNM built 14 new bridges and 32 tunnels, the longest of which, at 1.85 miles, represents the longest in all Latin America.

Over the remainder of the mountain division between Ciudad Mendoza and Paso del Macho, FNM is rebuilding or replacing the original line and adding a second main line. Near Fortin a spectacular new Metlac Viaduct has replaced the original viaduct with a double-track, six-span prestressed concrete structure that is 1410 feet long and stands 430 feet high. West of Paso del Macho FNM drilled a new 689-foot, double-track tunnel through the north end of Mount Atoyac. At one end this tunnel curves to join the unique Pensil Viaduct-Tunnel, a prestressed concrete viaduct linked to a 436-foot structure that resembles a giant shelf transversely anchored in the mountain. Concrete frames carry both floor beams that carry the double-track roadbed and arches that support a concrete vault over the roadbed, which protects the line from rock slides from the steep mountainside above. Nearby, a new ten-span prestressed concrete viaduct over the Atoyac and Chiquihuite rivers spans a total of 1627 feet.

Double-stack container trains now move over the much-improved route up Mexico's spectacular Maltrata Incline. Passengers can still enjoy the scenery on the journey once billed as the "railway trip of a thousand wonders" via a daily round trip that makes the journey between Mexico City and Vera Cruz during daylight hours.

GETTING THERE

While the Maltrata Incline is best seen from a train, Highway 150 and the Mexico City–Vera Cruz toll highway follow the same general alignment through the mountains.

The Southern Pacific's Tehachapi Pass Crossing of the Tehachapi Mountains (1876) near Bakersfield, California

Building south through California's San Joaquin Valley toward Los Angeles, the SP by 1875 had reached the very south end of the valley at Caliente. Here, at the toe of the Tehachapi Mountains, the railroad faced the formidable task of building across Tehachapi Pass to reach Mojave and Los Angeles. From Caliente to the summit of the pass, an air line distance of only about 16 miles, the railroad would have to climb some 2737 feet, more than half a mile, through a rugged and inhospitable terrain.

A War Department topographical party had surveyed possible routes through the pass in 1852. The SP chose the best of these and began surveys of its own in 1866, surveying lines at grades of 1.3, 2 and 2.2 percent before establishing the alignment that would be used. The route finally chosen in 1875 was the work of an innovative young assistant engineer, William Hood, who would go on to a long and successful career as the railroad's chief engineer.

Hood's alignment held the ruling grade on the north slope of the mountains to 2.2 percent. The line began at an elevation of 1291 feet at Caliente, crossing over two creeks and reversing direction through a great horseshoe curve to begin the ascent of the mountains. Reversing direction again, the line crossed Clear Creek Ravine to make its way back to the gorge of Tehachapi Creek. At Cliff, one could look down on Caliente, more than 700 feet below and scarcely a mile away as the crow flies, but 6 miles away by the railroad. The line was cut into a shelf in the side of the gorge, and one tunnel followed another as the line climbed up the rugged north slope of the Tehachapis. Altogether, 17 tunnels were required between the valley floor and the summit.

After crossing and recrossing Tehachapi Creek five times, the line turned away from the stream to reach the celebrated Tehachapi Loop at Walong. The Loop was

This view of the famous Loop at Walong in Tehachapi Pass soon after its completion shows the Southern Pacific's train 17, the *Los Angeles Limited,* on the Loop above Tunnel No. 9, from which it had emerged just moments before to begin the circuit of the Loop. A light engine had just followed the train out of the tunnel.—C. E. Watkins, James H. Harrison Collection, from Golden West Books.

conceived by Hood as a way to "make distance," enabling the railroad to maintain its 2.2 percent maximum grade by lengthening the line as it climbed. The 126-foot Tunnel No. 6 took the line under the side of a mountain ridge, and the line then followed a 10-degree curve through a great 360-degree loop, gaining 77 feet in elevation as it circled around a conical hill and back over itself in just the length of 80 cars.

There are legends about how Hood came up with the idea. One, for example, attributes his inspiration to the meanderings of a mule, taking the path of least resistance. In reality, it was simply good engineering. "It was just a common sense plan," Hood once said. In planning the line through the pass, Hood is said actually to have worked out the idea of the Loop first and then laid out the line above and below it.

Above the Loop the line followed Tehachapi Creek again toward the crest of the grade, 4028 feet above sea level. Here, the grade leveled out as the line reached the table land of the Tehachapi Valley and then began the descent along Cache Creek to the Mojave Desert. The grades were just as severe on the south slope of the pass, but here the terrain was open, and the line was easily laid through broad curves.

Construction through the pass began in mid-1875, and more than 3000 men and hundreds of horses and dump carts were soon at work on the grade. The workmen used some 600 kegs of powder every week to blast their way through the mountains. Four of the north slope tunnels were drilled through hard rock, while the remainder were tunneled through soft rock and lined with timber. By July of 1876 the line was operating all the way to Tehachapi Summit and all the way down the south slope to Mojave by early August.

Meanwhile, other SP construction crews had been building north from Los Angeles; and on September 8,

In this remarkable 1928 photograph of the Loop from above, six locomotives can be seen on an eastbound SP freight train climbing the Pass. Below the Loop in the distance, to the right, exhausts of still more locomotives mark the approach of another train climbing up through the pass.—Donald Duke Collection.

In the years of transition from steam to diesel power, four Electro-Motive F units headed the first section of eastbound SP train 806 through Tehachapi Pass in October 1950. The units had just cleared Tunnel No. 9 to begin the circuit of Southern Pacific's celebrated Loop.—William D. Middleton.

Rulers of the Southern Pacific's mountain grades for more than four decades were the railroad's extraordinary cab-forward articulated locomotives. One of them, Class AC-10 4-8-8-2 No. 4244, stormed up the north slope of Tehachapi Pass with the second section of eastbound train 806 in October 1950. The train was crossing Tehachapi Creek just below the celebrated Loop. Cut into the 58-car train behind the 2-8-8-4 were three helper engines for the long hard pull up the 2.2 percent grade to the summit.—William D. Middleton.

—Union Pacific.

William Hood (1846–1926) was born at Concord, New Hampshire. After serving in the Civil War, Hood entered Dartmouth College, where he earned a scientific degree in 1867. Like many an ambitious young man of the time, Hood then headed west to build railroads and soon began a career with the Southern Pacific that was to last for the next 54 years.

Hood worked almost everywhere on the expanding SP system. He helped build the CP's line over Donner Pass and east to Promontory. After completing the Tehachapi Pass line, he worked on the building of the Sunset Route. He became chief engineer for the CP in 1883 and was then chief engineer for SP's Pacific system during a 15-year period that saw the SP complete its Shasta Route into Oregon.

In 1900 Hood became chief engineer for the entire Southern Pacific Company, and over the next decade he directed completion of the Coast Line between Los Angeles and San Francisco and a massive Harriman improvement program that included such notable projects as the reconstruction and double-tracking of the CP's line over Donner Pass, construction of the Lucin Cut-Off across the Great Salt Lake (see page 42), and the building of the San Diego & Arizona. Hood retired from SP in 1921 and died at San Francisco five years later.

Ever since 1899, Southern Pacific has shared its line through Tehachapi Pass with the Santa Fe. On June 12, 1954, just a few days after the train's June 6 inaugural, westbound train 1, the *San Francisco Chief,* descended the north slope grade of the pass on its way to the San Joaquin Valley and the Bay Area.—William D. Middleton.

1876, the two lines met at Lang, in Soledad Canyon, for a golden spike ceremony that celebrated the linking of San Francisco and Los Angeles by rail.

The Santa Fe came to the San Joaquin Valley in 1898. There was talk of a new line across the mountains for the Santa Fe, but by 1899 the two railroads had arrived at a joint trackage agreement for the SP's line though the pass; and the two railroads have shared it ever since. As traffic of the two railroads steadily increased over the Tehachapi Mountains, the line soon became one of the busiest single track mountain railroads in the world.

Multiple helpers in both directions and successive generations of heavier motive power moved the growing tonnage over Tehachapi's 2.2 percent grades. Continuing improvements to the mountain crossing helped to increase capacity as traffic grew. In 1905 automatic block signals were installed over the entire route. Sidings were lengthened and new ones installed. By 1924 SP had completed double-tracking of large sections of the line. Periodically, there were proposals for entirely new lines, a long tunnel, or electrification across the mountains, but nothing this ambitious was ever undertaken.

Tehachapi Pass has always been a place of intense activity. In the record breaking month of September 1925, for example, the SP reported that 1131 trains had moved a total of 41,877 cars over the mountains. On the month's busiest day there were 110 separate train movements. During World War II, traffic carried by the two railroads through the pass reached a peak of 50.5 million tons in 1945.

Today, the SP's great mountain railroad follows much the same route over Tehachapi Pass that was laid down by William Hood a century and a quarter ago, and it remains a vital link between Northern and Southern California for the two giants of western railroading, UP and BNSF. Tehachapi Loop was designated a National Historic Civil Engineering Landmark by the American Society of Civil Engineers in 1998.

GETTING THERE

The SP's line through Tehachapi Pass is paralleled by California Route 58. The Loop is just off Woodford-Tehachapi Road about 3 miles east of the Keene exit.

The Canadian Pacific's Rogers Pass Crossing of the Selkirk Mountains (1885)

One of the most difficult tasks faced by the builders of the Canadian Pacific was the location of the best and most feasible route across the mountain ranges of British Columbia. Earlier surveys had identified at least three possible routes through the Rockies. The Selkirks were another matter, for no one had yet found a suitable route. They could be circumvented by following the Big Bend of the Columbia River to the north, but this would add several hundred miles to the route.

In 1881 the CP decided to take a southerly route through the mountains, even before it was sure a feasible alignment could be found. This was to be the job of Major A. B. Rogers, who had just been hired by James J.

The rugged terrain of the Canadian Pacific main line through the Selkirks is evident in this view dating to about 1884, showing newly laid track on the west slope of Rogers Pass about 1 mile from the summit. Visible in the distance is the Hermit Range.—O. B. Buell, Canadian Pacific Archives (Neg. A.4226).

One of the largest bridges on the entire Canadian Pacific was this enormous timber trestle and Howe truss structure built to carry the line over Mountain Creek in Rogers Pass. The bridge was 164 feet high and 1086 feet long, and contained over two million board feet of timber.—O. B. Buell, Canadian Railway Museum.

The fierce snowfalls of the Selkirks quickly convinced the CP that snowsheds would be required for reliable and safe operation over Rogers Pass. This view, dating to the late 1880s, shows construction of one of the 31 timber sheds that were built in the pass.—Glenbow Archives (Neg. NA-1753-21).

This view at the summit of the Selkirks looking south toward Rogers Pass, with Illecillewaet Glacier in the distance, dates to about 1888. At the right is Shed 17, while to its left is the summer track that was installed at William Van Horne's order so that passengers could view the splendid scenery of the pass.—A. B. Thom Photo, O. Lavallée Collection, Canadian Pacific Archives (Neg. A.1891).

Hill, one of the syndicate building the railroad, for just that task. After exploring several possible routes through the Rockies, Rogers recommended one through Kicking Horse Pass. Finding a route through the Selkirks was more difficult.

In 1865 Walter Moberly had surveyed several potential routes across the mountains. Seeking a direct route through the Selkirks, Moberly had explored the deep gorge of the Illecillewaet River. He found its north fork to be an impossible route. An assistant had found much more promising conditions along the river's east fork, but he failed to reach the summit to confirm that it was a feasible pass.

Rogers had read Moberly's journal, with its reference to the east fork of the Illecillewaet as a possible route through the Selkirks. In April 1881 he set out to the east from Kamloops, British Columbia, with his nephew and a party of ten Indians to explore the route. They arrived at the mouth of the Illecillewaet late in May and began a difficult climb up the deep canyon of the river through dense forests and deep snow to search for a pass. Finally they reached the summit and found what appeared to be a route down the eastern slope. But without sufficient supplies to continue, Rogers could not be sure.

The following year, Rogers approached the Selkirks from the east. Beginning where the Beaver River flowed into the Columbia, Rogers and his party struggled upward through the rugged terrain, finally reaching the same mountain meadow they had found the year before from the other side. It was July 24, 1882, and Rogers had found the pass that would bear his name and eventually carry the CP through the Selkirks.

CP construction crews working under James Ross reached the Columbia River by the end of the 1884 construction season. Early the next year Ross and his men began the daunting task of building the line across the Selkirks.

The line laid down by Rogers followed the west bank of the Beaver River, climbing through 1768 feet in 21.5 miles to reach the summit at 4351 feet, never exceeding a 2.2 percent maximum grade. Several enormous timber bridges were built to carry the line across the deep ravines of the streams that fed the Beaver. At Mountain Creek Ross's men built one of the largest bridges on the entire railroad, a structure 164 feet high and 1086 feet long that contained over two million board feet of timber. Stoney Creek was crossed by a great timber Howe Truss bridge 453 feet long and 292 feet high.

Even after the snowsheds were built, operation over Rogers Pass remained a difficult task for the CP's operating forces. This drawing by artist Melton Prior for *The Illustrated London News* of December 8, 1888, shows a typical winter scene in the Selkirks. The train is emerging from a snowshed at Ross Peak, near the Glacier House.—Author's collection.

Not long after the CP began operating over Rogers Pass, a westbound passenger train paused at Glacier station at the Illecillewaet Glacier. By 1916, completion of the 5-mile Connaught Tunnel ended the need for operation over this difficult section of line.—Canadian Pacific, David P. Morgan Memorial Library.

Just west of the summit the line followed a great loop, or elongated figure "8," in the valley of Fivemile Creek (now Loop Creek) to maintain the 2.2 percent maximum grade. Four major bridges across the creek and the Illecillewaet River were required for the loop. In all, the winding line down the west slope required a total of 2500 degrees of curvature, or seven complete circles. At Laurie, 14 miles west of the summit, Ross had to drill two tunnels, and then a third further down the slope at Albert Canyon. In just 45.5 miles from the summit the line dropped through 2857 feet, more than half a mile, by the time it reached Farwell (later Revelstoke) at the CP's second crossing of the Columbia River.

It had been the most difficult of construction seasons. When Ross reached the site to begin work in February, he found snow packed to a depth of 10 to 12 feet on either side of the summit, and 30 to 40 feet deep in some of the cuts. Snow slides were a frequent hazard. At one point, ten slides came down on the line in six days' time. Several men were buried in slides, and two were killed. In April the men went on strike when their pay fell months behind. In May there were forest fires, and the timber span across the Beaver River was damaged by fire. Later in the year, when construction had reached the west slope, severe rainstorms caused washouts; and high water in the Illecillewaet washed out a bridge.

But despite it all, the rails continued to advance to the west. The summit was reached on August 17, and by September 27 the railhead was another 24 miles to the west. Finally, on November 7, the CP's lines from east and west were joined in a simple last spike ceremony at Craigellachie in Eagle Pass. Canada's great transcontinental railroad was complete.

The heavy snows and the snow slides of Rogers Pass that had so troubled Ross were an equal threat to the CP's operating men. In that first winter, before the line was opened, as many as nine avalanches buried the

Albert B. Rogers (1829–1889) was a native of Orleans, Massachusetts. After studies at both Brown University and Yale, he earned an engineering degree in 1853. Rogers worked for a time on the Erie Canal before moving west to begin a railroad career. Except for a brief interlude as a major of cavalry during a Dakota Sioux uprising in 1862, Rogers spent the rest of his life in railroad location work.

—Glenbow Archives (Neg. NA-1949-1).

Rogers was engaged in the construction of two midwestern lines, and then in 1861 joined the engineering staff of the Chicago, Milwaukee & St. Paul. His ingenuity in discovering economical locations won him the name of "the railroad pathfinder." In 1881 this reputation brought him to the attention of James J. Hill and the Canadian Pacific.

If ever there was the quintessential railroad location engineer, it was Major A. B. Rogers. He was notorious for his ability to travel light, living off uncooked beans and biscuits. He was a tough and profane man who chewed tobacco incessantly. He drove his men hard and fed them poorly. Most quit as soon as they could, but others admired him greatly and went anywhere with him. Rogers was a driven man, and the challenge of locating the way across the Rockies was a chance at fame, rather than fortune. James J. Hill told him that a pass would be named for him if he found it, and it was. Instead of cashing the $5000 bonus check the CP gave him, he framed it and hung it on the wall.

After completing his work for the CP, Rogers was hired by Jim Hill again, this time as a locating engineer for the Great Northern. His engineering career came to an end when he was badly injured in a fall from his horse in 1887. He died two years later at the home of his brother in Waterville, Minnesota.

summer, CP president William Van Horne ordered special summer tracks laid alongside the snowsheds.

Even with the sheds, an annual snowfall of up to 40 feet remained a force to be reckoned with. An avalanche in 1899 destroyed the station at the summit, killing seven people. Another near Shed 17 in 1910 killed 62 men who were digging the line out from a previous slide. The line proved difficult and costly to maintain, while the long, steep grades on both sides of the pass represented a severe operating handicap as traffic grew.

Almost four decades after the line had opened, CP construction crews were ready to battle Rogers Pass again, this time with a tunnel through the 9491-foot-high Mount Macdonald that took the line under the summit. Drilled through in just 704 days, the 5-mile, double-track Connaught Tunnel was the longest in North America when it was completed in 1916. The tunnel lowered the grade through the pass by 540 feet, shortened the distance to Vancouver by 4 1/2 miles, and eliminated both the entire section of line with snowsheds and the Loop west of the summit.

In 1983, a century after James Ross completed the CP line across Rogers Pass, the railroad took on Rogers Pass once again to build still another record-breaking tunnel through Mount Macdonald that would take westbound traffic across the Selkirks at a still lower elevation (see page 136).

GETTING THERE

Rogers Pass lies about 40 miles east of Revelstoke on the Trans-Canada Highway, which generally parallels the CP's route through the Selkirks. The Rogers Pass National Historic Site is located near the summit in Glacier National Park.

The Great Northern's Marias Pass Crossing of the Rocky Mountains (1893) between Shelby and Whitefish, Montana

By the late 1880s James J. Hill had developed a formidable railroad empire that stretched all the way from St. Paul to Butte, Montana. But he had been outflanked by both the Canadian Pacific and the Northern Pacific, which had already completed routes to the Pacific. Early in 1886, Hill committed his railroads as well to a Pacific Coast terminal. "What we want is the best possible line, shortest distance, lowest grades and least possible curvature that we can build," he wrote in 1890. "We do not care enough for Rocky Mountain scenery to spend a large sum of money developing it."

As far west as Havre, Montana, Hill's Great Northern followed a direct route to the west between the 48th and 49th parallels, about 30 to 40 miles south of the

tracks at one point. At another, the line was buried under 39 feet of snow. Some avalanches were said to have exceeded a million tons of snow, roaring down the mountainside at more than 60 mph.

Even before through trains began running over the line in the spring of 1886, construction crews had begun the task of building 31 timber snowsheds, with an aggregate length of over 5 miles, on 16 miles of line over the summit. Almost 18 million board feet of sawn timber and more than a million feet of round timbers and piling were required, and the work took two years to complete. To make sure that passengers could still view the marvelous scenery and the great Illecillewaet Glacier during the

John F. Stevens's discovery of a favorable route through the Rockies at Marias Pass gave the Great Northern a route on the fringes of what would become Glacier National Park that offered its passengers unsurpassed scenery. In this view from the 1940s the GN's westbound *Empire Builder* is nearing Summit Station in the pass behind a 4-8-4 Northern.—Great Northern, David P. Morgan Memorial Library.

—Library of Congress (Neg. LC-USZ62-75399).

John Frank Stevens (1853–1943), while never formally trained in the field, was one of the most notable engineers of his time. Born at West Gardiner, Maine, Stevens was educated in public schools and the State Normal School. He began his engineering career in 1874 in a post with the Minneapolis city engineer. Two years later Stevens moved to Texas to take up the first of many railroad location and engineering assignments in the U.S. and Canada. Following his discovery of Marias Pass, Stevens went on to head the final location and construction work across Washington for the GN. He later became the railroad's chief engineer and then general manager, and he went to a similar post with the Rock Island during 1903–05.

In June 1905, Stevens took up one of the most challenging assignments of his career when President Theodore Roosevelt appointed him chief engineer of the Panama Canal. Stevens ran the canal project the way he had the Great Northern. He got the work going in good order, and led the fight for a locked rather than a sea level canal. But for reasons that he never made clear, Stevens abruptly resigned in 1907 and soon returned to railroad work. During 1917–23 he headed U.S. commissions that assisted the railway systems of Russia and China. One of his last engineering assignments was as a consultant to the Great Northern in planning its new Cascade Tunnel.

Stevens's remarkable career was recognized by such awards as the John Fritz Medal in 1925, the gold medal of the Franklin Institute in 1930, and the Hoover Medal of the American Society of Civil Engineers in 1938. He was an honorary member of the American Society of Civil Engineers and its president in 1927. John Stevens was 90 years old when he died at his home in Pinehurst, North Carolina, in 1943.

Canadian border. The most direct line to Puget Sound would be one that continued due west from Havre on what became known as the Assiniboine route. The only trouble with this route was that no one was certain there was a feasible pass for it through the Rockies.

As far back as the Lewis and Clark expedition of 1804–06, there had been surveys of routes across the Rockies, but all of them lay further south than Hill wanted to go. There was said to be another pass right where Jim Hill wanted it, but no one except Indians and perhaps a few trappers had ever seen it. This was Marias Pass at the headwaters of the Marias River, which flowed east from the Rockies to the Missouri.

The Marias River itself had been located by Lewis and Clark on their westward trip in 1805. On the return trip from the Pacific in 1806, Lewis had set out to explore the river and its headwaters. This was given up following an altercation with some Indians, and the reports of hostile Blackfeet that came out of the incident helped discourage further exploration of the area.

Major Isaac I. Stevens, who explored a possible northern route under the Pacific Railroad Survey of 1853, had also heard about Marias Pass. There were reports as early as 1810 that traders, accompanied by Indians, had crossed the Rockies here by a "wide defile of easy passage," and a map of the region published in 1840 showed a trail right about where Marias Pass would later be found. He had also heard of it from a Blackfoot chief named Little Dog, who had described it to Stevens when he stopped at Fort Benton on his way west in 1853. While Stevens explored further south during 1853–54, parties from his expedition explored the area of the elusive Marias Pass, but they failed to find it; the pass remained unexplored for another 30 years.

When James J. Hill was ready to build through the Rockies in 1889, he knew of Isaac Stevens's belief that there were probably better northern routes than those he had been able to explore. One of the railroad's reconnaissance and locating engineers had explored the Marias Pass area in 1887, but he had failed to find the summit. Hill was determined to make one more effort to find and explore the pass.

The man given the task was John F. Stevens, a remarkable locating and construction engineer, who had just joined the Great Northern. The story of Stevens's

Across Great Mountains   **97**

Not long after the 5400-hp diesel was delivered in 1943, a Great Northern photographer recorded the railroad's four-unit Electro-Motive FT No. 410 climbing the west slope of Marias Pass near Blacktail, Montana, with a 106-car train of refrigerator cars. A three-unit diesel pusher helped the train up the 1.8 percent grade toward Summit.—Great Northern, David P. Morgan Memorial Library.

search for Marias Pass has become one of the legends of railroad location.

It was December, far too late in the year to be seeking a pass through the Rocky Mountains, but Stevens set out anyway. As he wrote years later, "I was 36 years of age, my middle name was 'nerve,' and as I had no notion of lying idle if any work of an engineering nature was in sight, I at once accepted the invitation." Accompanied by one assistant, Stevens headed west from Fort Assiniboine, near Havre, with a mule team and driver, and a saddle-horse.

Struggling through a succession of blizzards, Stevens and his party finally reached the Badger Creek Indian agency, 160 miles from Fort Assiniboine. There, they were told it was impossible to go any further by team. Stevens succeeded in covering a few more miles, but his assistant refused to continue. Stevens then tried to hire some of the Blackfoot Indians to guide him, but none would accompany him, referring to a "bad spirit" they dared not meet at the head of the river in the mountains. He finally hired a Flathead Indian for the task, but the man had never been near the pass, as it turned out, and was of little help to Stevens.

Leaving the team and his assistant behind, Stevens and his Indian guide set out toward the pass on improvised rawhide snowshoes, each carrying a pair of blankets and a very limited supply of food. The two men climbed upward through two to four feet of snow as far as a point called "False Summit." Beyond this point the snow grew deeper and the weather colder. "My Indian played out on me at the small creek just west of false summit," wrote Stevens, "and I built him a fire and pushed on all by my lonesome."

Whether by his pathfinding skill or plain luck, after several attempts Stevens managed to walk right into the pass. A little creek, later named Summit Creek, had led him to the Continental Divide. Beyond, another stream flowed west toward the Middle Branch of the Flathead

River. On December 11, 1889, more than 80 years after Meriwether Lewis had first sought it in vain, John Stevens had found Marias Pass.

After reaching the summit, Stevens continued beyond to make sure that the drainage was really to the west and then returned to the summit to camp for the night. It proved impossible to build a fire and keep it going, so Stevens tramped out a track about a hundred yards long and walked back and forth all night to keep from freezing to death. Later, he learned that even on the plains at the Indian agency the temperature had dropped to 36 degrees below zero that night. What the temperature had been 1500 feet further up where he spent the night, said Stevens, "the good Lord only knows but the mosquitoes didn't bother me."

"It was a strenuous trip and I am frank to say," he recalled years later, "that none but an unusually strong man, as I was in those days, could have made it."

A little over a month after Stevens had located the summit of the pass from the east, locating engineer C. F. B. Haskell and two men went in on horseback and on foot from Demersville, at the head of Flathead Lake near what is now Kalispell, to explore the western approach to the pass. Haskell and his men encountered temperatures as low as 44 degrees below zero and snow 6 feet deep, but they successfully located a line down the west slope of the mountains

Stevens's discovery of Marias Pass gave the GN a route that was more than a hundred miles shorter than any other it could have taken, crossing the Continental Divide at an elevation of only 5213 feet without a tunnel. From the east the line followed the Marias and its tributaries into the pass on a 1 percent ruling grade. Beyond the summit the line dropped down to the west on 14 miles of 1.8 percent ruling grade through John F. Stevens Canyon and then followed the Flathead, Kootenai, and Pend Oreille rivers on grades of 0.7 and 0.8 percent. Except for the Canadian National's transcontinental main line through Yellowhead Pass far to the north, which crossed the Divide at 3717 feet, the GN had a lower crossing of the Divide and better grades than any of the other northern lines.

"That discovery," wrote Burlington president Ralph Budd of Marias Pass many years later, "changed the Great Northern Railway from a circuitous railroad into a very direct transcontinental line, with very favorable grades and alignment on both easterly and westerly approaches."

Construction of the GN's Pacific extension west from Pacific Junction had begun late in 1889, even before the final route through the Rockies had been determined. Construction crews built through Marias Pass from both east and west. Materials and supplies on the east side advanced with the railroad, while those for the west side were carried from an NP railhead by wagon and steamboat to a construction base camp at Demersville.

Hill pushed his men relentlessly, and the trains were running through Marias Pass to the west by 1892. The best and shortest route to the Northwest when it was completed in 1893, Jim Hill's road to the Pacific remains so today as a principal route of Burlington Northern & Santa Fe.

GETTING THERE

U.S. 2 follows the route of the GN main line all the way through Marias Pass. The Coy–Del Grosso book listed in the bibliography is an excellent guide to specific vantage points.

The Chihuahua Pacific's Crossing of the Sierra Madre Occidental (1961) in Chihuahua and Sinaloa, Mexico

Few great North American railway building projects took longer to realize than the dream of a line across Texas and Mexico to a new port on the Gulf of California, linking mid-America with the Orient via the shortest and most direct route to the Pacific Ocean.

The idea was first advanced in the 1880s by Albert K. Owen, who proposed an "International Air Line" that would link Asia to Europe via the U.S. and Mexico through ports at Topolobampo, Sinaloa, on the Gulf of California, and Norfolk, Virginia. Owen was unable to finance any construction, but at the turn of the century

Building the Chihuahua Pacific over the rugged terrain of the Sierra Madre Occidental required the erection of 37 bridges and drilling or construction of 86 tunnels, and one follows another in rapid succession in the steep canyons of the *Barrancas del Cobre*—Copper Canyon—region. One of each was in sight as the northbound tri-weekly *Vista Train* made its way up the west slope of the Sierra Madre on December 28, 1975.—William D. Middleton.

Across Great Mountains

Over 1018 feet long, and standing 355 feet above the water, the Rio Chinipas bridge is the highest on the Chihuahua Pacific. A High Iron Travel special train paused on the massive structure in November 1996.—J. W. Swanberg.

Mexican businessman Enrique Creel was able to build the first segment of a line from Chihuahua across the Sierra Madre Occidental to the Gulf. Financier Arthur Stilwell soon joined forces with Creel to promote the Kansas City, Mexico & Orient Railway to complete the line. But after building a stretch of line in Kansas and Texas, and isolated segments in Mexico, the Orient went bankrupt, and the dream of a new route to the Pacific went unfulfilled for another 50 years.

The project came alive again in 1953, when the Mexican government organized the Ferrocarril Chihuahua al Pacifico to complete a line across the Sierra Madre. Track had already been completed up the gentler eastern slope of the mountains to Creel, and 73 miles inland from the Gulf port of Topolobampo to Hornillos, but there remained a 158-mile gap through the most difficult part of the rugged Sierra Madre.

To cross the mountains the line would have to climb to a maximum elevation of 8071 feet at Los Ojitos, before beginning the descent to sea level through the steep slopes and deep canyons of the Sierra Madre. Chief among these great canyons is the celebrated Barranca del Cobre, or "Copper Canyon," which surpasses even the Grand Canyon in scale.

The engineers surveyed a route through this rugged terrain with grades of no more 2.5 percent, but only with great difficulty. Great loops and horseshoe curves were required, and curvature on the line reached a maximum of 9 degrees. At some locations as many as three distinct levels of line could be seen from a single point. At El Lazo, between Creel and La Laja, the line coiled back on itself to form a complete loop in a distance of only a mile and a half, the upper part of the loop crossing a bridge at the point of intersection while the lower part passed through a tunnel. At Temoris, the line reversed direction twice to climb high above the valley of the Rio Septentrion, first reversing direction on a curved bridge across the river and then again in the 3,074-foot curved La Pera tunnel as the track climbed up the wall of the canyon.

There were 37 bridges of steel or prestressed concrete. The longest was a 1638-foot crossing of the Rio Fuerte made up of seven 108-foot steel girder spans and a 767-foot, three-span continuous steel truss main river crossing. The line's tallest bridge was a similar continuous steel truss crossing of the Rio Chinipas that stood 355 feet high.

Most of the line's 86 tunnels were drilled and blasted through rock, but more than 20 were "false" tunnels, created by erecting earth-covered concrete arches above deep cuts at locations where slides were common. El Descanso, the line's longest tunnel, was more than a mile long.

The Mexican government began work on the line in 1954. Over the next seven years the railroad builders

El Divisadero is a mandatory stop for every passenger train on the Chihuahua Pacific route over the Sierra Madre. Here, passengers disembark to view the spectacular Copper Canyon from an overlook some 4135 feet above the Rio Urique, and the local Tarahumara Indians do a brisk business at trackside souvenir shops and food stalls. A ten-car *Vista Train* paused there in December 1975.—William D. Middleton.

One of the Chihuahua Pacific's most spectacular features comes at Temoris, where the line crosses the Rio Septentrion to reverse direction, and then reverses direction again in a tunnel as it climbs up from the river valley. Powered by a single Electro-Motive unit, eastbound train 73, the daily Los Mochis–Chihuahua *Servicio Estrella* first class train, crossed the curved Santa Barbara Bridge over the Septentrion in June 1998 to begin the first of the two horseshoe curves that carry the line up the steep canyon wall.—William D. Middleton.

moved more than 19 million cubic yards of earth and rock, blasted more than 10 miles of tunnel through the rock of the Sierra Madre, and erected some 4 miles of bridges to complete the line. On November 24, 1961, President Adolfo Lopez Mateos officially opened the line to Topolobampo.

The major new route to the Orient envisioned by Arthur Stilwell and others never materialized, but the Chihuahua Pacific soon became an important artery of commerce for northern Mexico and the route for what is arguably the most spectacular train trip on the North American continent. The line became part of the Mexican National system in 1987, and has been operated as part of the privatized Ferrrocaril Mexicano since early 1998.

GETTING THERE

The most spectacular mountain sections of the Chihuahua Pacific are best seen from the train. But while much of the line has long been virtually inaccessible by road, this is fast changing. From a junction with east-west Hermosillo-Chihuahua Highway 16 a few miles west of La Junta, a paved road now generally follows the east end of the line south through Creel to the Copper Canyon overlook at Divisadero. One can follow the route another 30 miles or so by very primitive unpaved roads to reach the spectacular double loops at Temoris. From the west end a paved road northeast from Los Mochis to La Fuerte and Choix generally follows the railroad, but this route ends well before the line reaches its most spectacular sections. Excellent maps of the Sierra Tarahumara or Copper Canyon region and the most interesting sections of the railroad are available from the International Map Co. at the University of Texas at El Paso, Box 400, El Paso, TX 79968-0400.

Across Great Mountains

This imaginative drawing from *Scientific American* for September 13, 1890, illustrated the historic moment when the shields for the St. Clair River railway tunnel met beneath the river on August 30, 1890.—Author's collection.

3
Railroads below Ground

The art of tunneling has been practiced since ancient times. Early civilizations built a variety of underground shafts, temples, tombs, aqueducts and the like, with some early Egyptian subterranean tombs dating as far back as 1500 B.C. The Romans were the greatest early tunnelers of all, having built several road and aqueduct tunnels a mile or more in length through hard rock. One Roman tunnel completed by Emperor Claudius in 52 A.D. to drain a lake east of Rome was three and a half miles long and took 11 years to build.

These early tunnels were all the more remarkable for having been built in the absence of either machinery or explosives. Rock was most often removed by hammer and chisel. The Egyptians used channeling and wedging, a method in which wooden wedges were driven into channels cut into the rock, which was then broken out by the swelling action of the wedges when soaked with water. Still another method was to heat the rock to a very high temperature with fires and then to quench it with water, causing the rock to shatter from the sudden change in temperature. These primitive tunneling methods remained in use until the seventeenth century, when German miners began to drill and blast with gunpowder.

The first railroad tunnels were built in France and England between 1826 and 1830. The first in North America was at Staple Bend along the Conemaugh River on the Allegheny Portage Railroad near Johnstown, Pennsylvania. This was a 901-foot tunnel with an arch section 19 feet high and 20 feet wide that was driven through a slate hill during 1831–32. The tunnel remained in service less than a quarter century before the entire Portage Railroad was rendered obsolete by completion of the Pennsylvania Railroad's all-rail route across the Alleghenies. The tunnel is still there, however, and this enduring artifact of early railroad construction is now part of the Allegheny Portage Railroad National Historic Site, which was placed on the National Register

America's first railroad tunnel was this 901-foot bore through a slate hill at Staple Bend on Pennsylvania's Allegheny Portage Railroad. The historic tunnel, which by then had been abandoned for more than 30 years, was illustrated for the June 29, 1889, issue of *Engineering News*.—Author's collection.

Among the very oldest of North American railroad tunnels is the 300-foot Bundy Hill Tunnel at Taftville, Connecticut, which was blasted through the hard rock hill by the Norwich & Worcester in 1837. The venerable tunnel remains in service today on what is now the Providence & Worcester. The railroad's diesel locomotive No. 2009 negotiated the tunnel with a 13-car southbound train NR-2 on January 2, 1998. —J. W. Swanberg.

of Historic Places in 1967, and was designated a National Historic Civil Engineering Landmark in 1987.

Among other early railroad tunnels in the U.S. were the Norwich & Worcester's tunnel at Tafts, Connecticut, and the Philadelphia & Reading's Black Rock tunnel near Phoenixville, Pennsylvania, both completed in 1837 and both still in service today. These were soon followed by many more, and by 1850 some 48 railroad tunnels had been completed in the U.S.

These early tunnels were typically driven through hard rock by laboriously drilling holes with hammer and chisel and then blasting with black powder. Removal of the excavated material, or mucking, was usually accomplished by means of horses and wagons or small rail cars. Sometimes the surrounding rock was firm and stable enough that no additional support was required, but more often some form of timber or masonry lining was required.

For as long as tunneling continued to be done this way, it was an exceedingly costly and time-consuming process. In drilling the Virginia Central's Blue Ridge Tunnel at Rockfish Gap, Virginia (see page 109), which proved to be the longest ever completed by hand drilling and black powder blasting, the tunnelers averaged just 26.5 feet a month at a cost of $114 a foot. Tunnelers using the same primitive methods struggled ineffectually for some 15 years to drill the 4.75-mile Hoosac Tunnel in Massachusetts before the introduction of improved tunneling methods finally brought the project to a close (see page 111).

The successful completion of the Hoosac and other great tunnels of the late nineteenth century, such as the even longer Mont Cenis, Gotthard and Arlberg tunnels in Europe, clearly required better tunneling methods, and they came in the form of compressed air rock drills, nitroglycerin and dynamite that were developed during the 1860s. The new compressed air drills mounted in batteries on a moving drill jumbo or carriage could complete the blast holes in a fraction of the time required for manual drilling; and the new explosives were nearly half again as effective as black powder while their smokeless qualities eliminated the severe problems associated with the toxic fumes from powder.

In traditional tunneling methods, the full cross section of larger tunnels, such as those for railroads, was usually not excavated at one time. The most common method of excavation for hard rock tunnels was the heading and bench method. A heading, or drift, was a smaller segment drilled within the cross section of the tunnel. In the heading and bench method a heading the full width of the tunnel was driven at the top of the tunnel section, while the lower portion, or bench, was excavated later. Sometimes, too, hard rock tunnels were driven in a series of headings. Quite commonly, tunnels constructed in this manner would be driven with as many as six to eight separate drifts or headings. This had the advantage of allowing work to proceed at a number

Although it was completed several months later than the Taftville tunnel in Connecticut, the Reading's 1932-foot Black Rock Tunnel near Phoenixville, Pennsylvania, has been in service longer than any other North American rail tunnel. A westbound Conrail freight train entered the tunnel in September 1985.—Herbert H. Harwood, Jr.

The introduction of compressed air rock drills, together with nitroglycerin and dynamite, enabled nineteenth-century tunnelers to successfully complete such long tunnels as the 4.75-mile Hoosac or the even longer 8-mile Mount Cenis, between Italy and Switzerland. This drawing from *Harper's New Monthly Magazine* for July 1871 shows a battery of the new compressed air drills at work in the Mount Cenis Tunnel.—Author's collection.

of different points along the axis of the tunnel, and it facilitated the placement of timber supports when these were needed.

Another approach often used for long rock tunnels was the center heading method, in which a heading or drift at the center of the tunnel cross section was driven through from portal to portal and the tunnel then was enlarged to its full cross section by radial drilling from the center heading.

For very long tunnels, a parallel heading was sometimes excavated first to establish a smaller pilot or pioneer tunnel. Transverse headings were then driven over to the main tunnel axis, allowing tunnel excavation to proceed in both directions from several locations. The pilot tunnel was used to transport materials and workers to and from the drilling faces and as a drainage and ventilation tunnel.

When support of the tunnel section was required to prevent caving-in of the tunnel roof or sides, it was usually provided in the form of temporary timbering and strutting, as it was usually called, that was installed in stages as excavation was completed. This temporary support was then replaced by a permanent tunnel lining. This was sometimes of timber, but more often was of brick or—later—concrete.

Tunnels in soft or loose material required different methods of construction and always required both a temporary supporting structure and some form of permanent tunnel lining. While small tunnels might be constructed in a single heading, larger railroad tunnels

These drawings show how timber strutting was placed for typical stages of work in the Austrian method of tunneling. Top and bottom headings were drilled first, and then successively enlarged until the full tunnel section was excavated.—Author's collection.

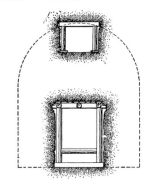
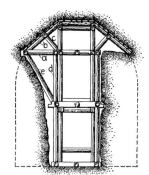
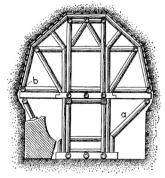

FIGS. 88 to 90.—Sketches Showing Construction of Strutting, Austrian Method.

Railroads below Ground

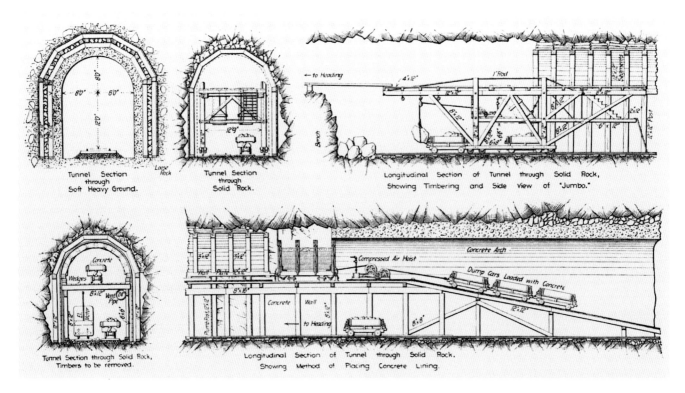

A series of drawings from *Engineering News* for January 10, 1901, show how the Great Northern's first Cascade Tunnel was drilled and lined.—Author's collection.

were typically excavated by a variety of combinations of multiple headings, followed by the installation of temporary timber supports and then the construction of permanent masonry or concrete lining.

Most of these soft ground tunneling methods took the name of their country of origin. The American method, for example, started with excavation of a top heading at the center, or crown, of the tunnel arch. This was then widened on each side, excavated down each side to the invert level, and finally excavated at the center. Another was the Austrian method, in which excavation began with a full-height center heading, followed by widening of the tunnel arch to its full width, and then excavation of the lower section at each side of the tunnel. Still other variations were known as the Belgian, English, German, and Italian methods. In every case, temporary timbering was installed at each stage of the excavation.

Tunneling under water presented an entirely different set of problems. The combination of hydraulic pressure with the soft or fluid material typically found in the bed of streams made it impossible to use ordinary tunneling methods, and there had been few attempts at subaqueous tunneling before the early nineteenth century.

The first important underwater tunnel was driven by British engineer Marc Isambard Brunel under the Thames River, between the Rotherhithe and Wapping sections of London, over an 18-year period beginning in 1825. To tunnel through the soft water-bearing clay of the river bed, Brunel designed a novel device called a "shield," which supported the soft material as the excavation progressed. Brunel's rectangular shield was made of cast iron, with shelves at the top, bottom, and sides which supported the roof, floor, and walls of the tunnel until a permanent brick tunnel lining was installed behind the shield. The working face of the tunnel was supported by a number of "breasting boards" which were held in place by screws bearing against the framework of the shield. One at a time these breasting boards were removed, the material behind them excavated, and the board then replaced in the advanced position. Once this had been completed across the full face of the tunnel, the shield was advanced by screw jacks bearing against the completed masonry tunnel lining behind the shield, and the process was repeated.

The project proved both difficult and costly, and the work was interrupted on a number of occasions when the river broke through into the tunnel or Brunel's finances ran out; but it did demonstrate for the first time an effective method for underwater tunneling. The Tower Subway, a second Thames tunnel completed in 1869 by James Henry Greathead, used a circular wrought iron shield based upon a design patented by civil engineer Peter W. Barlow. Much simpler than the Brunel shield, it had a sharpened circular ring in front that helped to penetrate the ground. Typically, tunnelers worked in front of the shield, excavating the clay of the river bed and passing it back through a doorway in the front of the shield. Another important advance of the Greathead tunnel was the use of segmental cast iron tunnel lining, which could be bolted in place much more rapidly than a masonry lining as the shield advanced. While it had taken Brunel 18 years to tunnel under the Thames, Greathead made it in less than a year and at far less cost.

While Greathead was tunneling under the Thames, a somewhat similar tunneling project was being completed at New York by Alfred Ely Beach, a noted inventor and the editor and publisher of *Scientific American*. Beach used a similar shield, circular in cross section, to drive a 312-foot subway under Broadway that was used to demonstrate Beach's design for a pneumatic subway. An important advance over the Greathead-Barlow shield was the use of hydraulic rams, rather than screw jacks, to advance the shield.

Another New York tunneling project a few years later introduced the use of compressed air for underwater tunneling. This was a project to tunnel under the Hudson River developed by DeWitt Clinton Haskin, a Californian who had made his fortune in western railroad construction and Utah mining. Haskin had become interested in the use of compressed air caissons for underwater construction after James Eads had successfully adopted the technique for the piers of the St. Louis Bridge. In 1874 he patented a tunneling method that eliminated the use of a shield, using only air pressure to keep water out of the tunnel until the permanent lining was installed. The use of air pressure for underwater tunneling was an important advance, but the elimination of the shield was not. Haskin's company struggled for nearly 20 years to finish the tunnel. When the project was finally completed in 1908 by another company, it was with a more conventional shield technology.

By the late nineteenth century these early efforts had formed the basis for a well-developed underwater tunneling technology that combined the use of compressed air ground support during construction, a movable shield advanced by hydraulic jacks, and a segmental cast-iron tunnel lining. Its first major application to a railroad tunnel was for the Grand Trunk Railway's St. Clair River tunnel between Sarnia, Ontario, and Port Huron, Michigan, which opened in 1891 (see page 114). Later shield tunneling projects included the Pennsylvania Railroad's Hudson and East river tubes at New York and a number of underwater subway, highway and utility tunnels there and elsewhere. Shield tunneling was also widely used for other types of soft ground tunneling, as well as underwater work.

Shield tunneling under water was always an exacting, risky business. A compressed air "blow out" into the river bottom, which could send water cascading into the tunnel, was an ever-present danger in soft or silty soils. Tunnelers had to enter and leave the pressurized tunnel through air locks, and the length of time they could work under pressure was limited. Caisson disease, commonly known as the "bends," was a hazard of working under air pressure.

An alternate sunken-tube underwater tunneling method was pioneered by the Michigan Central's Detroit River tunnel completed in 1910 between Windsor, Ontario, and Detroit (see page 127). A trench was first dredged in the bed of the river along the alignment of the tunnel, and steel tunnel sections fabricated elsewhere were then floated into place over the trench and lowered into position. Using the steel sections as a form, exterior concrete was then placed around the tunnel from a barge above the trench. Once all the tubes and exterior concrete were in place, wooden plugs at the ends of each tube section were removed and the tunnel was lined with concrete. This procedure had the advantage of eliminating much of the dangerous and costly underwater work associated with shield tunneling and, wherever conditions permit, has been widely used for both rail and highway tunnels ever since.

The development of new materials and new types of heavy tunneling machinery has brought substantial change to tunneling technology over the last half century. Hard rock tunnels are often drilled by full-face tunneling, in which the entire tunnel cross section is excavated at one time with the use of high-speed drills

Tunnel workers are shown with one of the dreadnought drills used during construction of the Canadian Pacific's Connaught Tunnel in Rogers Pass, British Columbia, during 1913–16. —Glenbow Archives (Neg. NA-4598-4).

Tunneling with a hydraulic shield is depicted in this drawing from the November 1, 1890, *Scientific American* of work in the Hudson River Tunnel construction. As the shield is moved ahead by hydraulic jacks, material from the river bottom enters the shield through doors and is removed by the tunnel workers. This particular tunnel, begun by DeWitt Clinton Haskin in 1874, was finally completed as one of the Hudson & Manhattan tubes that linked New York and New Jersey in 1908.—Author's collection.

mounted on huge track or rubber wheel mounted jumbo frames. Mucking, the removal of excavated material, is done with large mechanized loaders, and the material is removed by trucks or mine cars. Prefabricated steel sections have largely taken the place of temporary timber supports, while precast concrete sections have often been used in place of segmental cast-iron tunnel liners.

The greatest change of all, however, has come through the development of enormous tunnel-boring machines, which include the machinery for driving the machine forward and for placing a tunnel lining as well as for excavation. For hard rock tunnels these machines are typically fitted with a rotating head on which cutting tools are mounted, while hydraulic jacks force the cutting head against the rock face under heavy pressure. Tunneling machines for soft ground are sometimes fitted with some kind of a rotating cutter head within a shield, while a backhoe type is more effective in some soils.

The idea of these tunneling machines is nothing new. One had been tried—and proved a failure—as early as 1852 on the Hoosac Tunnel, while a 1914 textbook on tunneling identified no less than 46 patents that had been issued to that time for some sort of tunneling machine. But it was not until relatively recent years that reliable and effective machines were finally developed.

Where conditions are suitable for their use, these machines have greatly reduced the time required for tunneling. Portions of the Canadian Pacific's new 9.1-mile Rogers Pass Tunnel, completed in 1988 (see page 136), were drilled by the top heading and bench method, in which the heading was drilled by a tunnel boring machine. This huge machine proved capable of advancing the heading at rates of over 200 feet a day in some of the softer rock encountered.

The Canadian National's new St. Clair River tunnel, completed in 1995 (see page 117), was drilled with an enormous soft-soil tunnel boring machine that was fitted with spade-type and ripper teeth and twin disc cutters, while the shield was pushed forward by hydraulic cylinders to cut through the clay of the river bottom. Guided by lasers and satellite navigation, the enormous machine advanced under the river at an average rate of about 26 feet a day, installing a segmented precast reinforced concrete tunnel lining as it went.

Similar machines have become commonplace in modern subway construction. Those being used for the Los Angeles County Metropolitan Transit Authority's Red Line subway, for example, are 200-ton, 185-foot-long laser-guided machines with a 22-foot diameter shield. The shield is advanced by 16 100-ton hydraulic rams that can exert a force of 6 million pounds on the cutting edge. A backhoe excavator in the shield can remove about 850 cubic yards daily, allowing the machine to advance at a rate of anywhere from 60 to 100

Typical of modern tunnel-boring machines is this big machine, known as "the mole," used to drill 5 miles of the east end of the Canadian Pacific's Mount Macdonald Tunnel in Rogers Pass during 1984–87. The 22-foot 4-inch diameter machine drilled a top heading at rates of up to 174 feet a day. The bench below this top heading was then removed by drilling and blasting.—Canadian Pacific.

feet a day. A conveyor system carries muck to the rear, while precast concrete tunnel lining segments are wedged into place by an attachment on the rear of the shield.

While drilling the tunnel itself has always been the most difficult part of tunneling, tunnel builders have always faced other challenges as well. Whether beneath a river or in the heart of a mountain, water intrusion is almost always a problem; keeping it out in the first place, or draining it from a tunnel, can require some innovative engineering and construction work. Adequate ventilation of tunnels is always a requirement; and the longer the tunnel the harder it is to achieve. Indeed, the problem of clearing locomotive smoke and gases remained a major barrier to the use of long or underwater tunnels until electric operation became feasible. The problems have become more manageable with diesel motive power, but a long, heavy traffic tunnel still requires a substantial, high capacity ventilation system.

The Blue Ridge Railroad's Blue Ridge Tunnel (1858) at Rockfish Gap, Virginia

At mid-nineteenth century, early Virginia railroads had begun to build west to reach across the rugged ridges of the Appalachians to the western part of the state and the Ohio River. The Louisa Railroad was already building west from Charlottesville toward Rockfish Gap in the Blue Ridge Mountains. But building across the mountains would be a daunting task for the railroad, and its builders appealed to the state for help. This came in 1849 when the state formed the Blue Ridge Railroad to build a line over the mountains. This would extend from a connection with the Louisa at Mechums River, west of Charlottesville, across the Blue Ridge to Waynesboro, where the Louisa would continue to the west.

Claudius Crozet, an eminent French-born engineer, was appointed chief engineer of the new railroad, and no one could have been better qualified for the task. A decade earlier, as the state's principal engineer, Crozet had made surveys for a railroad route across the Blue Ridge and had determined that the best route was the one through Rockfish Gap.

The line planned by Crozet would ascend the east slope of the mountains to pass under the crest in a long summit tunnel, continuing to climb on a 1.3 percent grade until it reached the west portal. This 4264-foot Blue Ridge Tunnel would be the longest railroad tunnel yet built anywhere, and it would take Claudius Crozet and his men almost nine years to complete it.

Crozet and his assistants were at work by the summer of 1849, locating the line of the tunnel and its

A view of the west portal of Blue Ridge Tunnel before it was replaced by a new parallel tunnel in 1942. Visible just above the tunnel mouth is the commemorative stone tablet placed there when the historic tunnel was completed in 1858. —T. O'Meara, C&O, David P. Morgan Memorial Library.

A century and a half after it was opened to traffic, Claudius Crozet's Blue Ridge Tunnel remains largely intact, even though it has been idle for more than 50 years. This July 1998 view shows the egg-shaped, brick-lined section that was required because of badly fissured rock at the tunnel's west end.—William D. Middleton.

A Historic American Engineering Record photograph taken on the east slope of the Blue Ridge shows the east portals of the original Blue Ridge Tunnel, to the right, while the portal of the new tunnel completed in 1942 is to the left and below.—Library of Congress (Neg. HAER-VA-5-1).

approaches. It was to be excavated in a straight line between portals located in deep cuts, passing about 700 feet below the peak of the mountain. Designed for a single track, the tunnel would have an elliptical cross section, 16 feet wide and 20 feet high above the level of the rails.

Contractors excavated the tunnel from both ends. At the east end the tunnelers soon encountered extremely hard material, which quickly wore out drills and slowed progress to as little as 19 feet a month. Very little hard rock was found at the west end and the work went rapidly. However, the roof was full of fissures and frequent rock falls made timbering necessary. Before the tunnel was complete, some 485 feet in from the west portal was arched with brick. In 1854, the tunnelers from the west heading encountered more of the badly fissured rock, and it was necessary to timber and then brick arch another 272-foot section. Altogether, almost 800 feet of the tunnel had to be arched with brick.

Smoke from the black powder blasting was a continuing problem for the tunnel workers until Crozet devised a ventilation system that was powered by mules on a treadmill. Water seepage into the tunnel was another problem. At the east end the water flowed out naturally because of the descending grade, but at the west end it had to be pumped out.

Claudius Crozet (1790–1864) was born at Villefranche, France, and studied engineering at the Ecole Impérial Polytechnique in Paris. As an officer in Napoleon's army, Crozet was taken prisoner in Russia, finally returning to France to serve again under Napoleon until the defeat at Waterloo. Crozet's decision to emigrate to the United States in 1816 brought a remarkable engineering talent to Virginia.

—Smithsonian Institution.

Claudius Crozet served his adopted country as both an educator and an engineer. He was a professor of engineering at West Point from 1816 until 1823. He later headed schools in Virginia and Louisiana and was one of the founders of the Virginia Military Institute; but after he left West Point, Crozet's achievements were largely as an engineer. Twice, he served as the principal engineer of Virginia, and he spent several years in a similar position with Louisiana.

After completing the Blue Ridge Railroad in 1857, Crozet took up an engineering position on construction of the aqueduct at Washington. He returned to the field in 1860 as chief engineer of the Virginia & Kentucky until construction was halted at the outbreak of the Civil War. Before the war was over, he died at Midlothian, the country home of his daughter and son-in-law, just a few weeks past his 74th birthday.

The problems were unending. The contractors soon found they were losing money and the contract was renegotiated, more than doubling the excavation costs from the initial estimates. Several times the Irish tunnel workers went on strike. There were frequent accidents and a number of deaths from the hazardous work. In 1854 an outbreak of cholera in the shanty towns where the tunnel workers lived caused several dozen deaths. Crozet dealt with frequent complaints about the slowness of the work from members of the public and the General Assembly, and he worried about the state appropriating the money needed to keep the project going.

Despite all the problems, the tunnel workers continued to advance from the two headings. The two crews finally holed through the long tunnel on Christmas Day, 1856. There was still much to do, and it took all of 1857 to complete excavation of the tunnel floor for the installation of track and to complete the arching.

Trains began operating over Rockfish Gap long before the tunnel was complete. The Virginia Central, which had taken over from the Louisa Railroad, had completed a temporary line over the pass on the roadbed of the Blue Ridge Railroad and had crossed over the summit on switchbacks. Trains began running over this temporary line in the spring of 1854.

On the night of April 12, 1858, the completed tunnel was finally linked to the tracks on either side. The next morning the down mail train operated on time through the world's longest railroad tunnel, completing the passage in just six minutes.

The tunnel under the Blue Ridge was a remarkable achievement for Claudius Crozet, and it was never surpassed using the hand drilling and black powder blasting methods that he had employed. A few years later tunnelers would try to do it to drill a tunnel of almost 5 miles under Hoosac Mountain in Massachusetts, but success would elude them until better materials and methods became available.

The Blue Ridge Railroad and its tunnel became part of the Chesapeake & Ohio a few years later; and the tunnel served the C&O well for almost 85 years, before the dimensions of ever larger motive power finally exceeded the limits of Crozet's elliptical arches. In 1942 the C&O drilled a new Blue Ridge Tunnel under Rockfish Gap, slightly below the original tunnel on a new alignment. Using compressed air drills mounted on a big drill jumbo, the modern tunnelers bored their way through the mountain at an average rate of 8 feet each eight-hour shift, completing in less than a year what it had taken Crozet eight years to do.

The old tunnel is still there under Rockfish Gap. Sections of the brick arch have begun to fall in, but an intrepid hiker can still make the 4264-foot passage under the mountain. The tunnel was designated a National Historic Civil Engineering Landmark by the American Society of Civil Engineers in 1976.

GETTING THERE

Both I-64 and U.S. 250 pass through Rockfish Gap. The original Crozet tunnel is just north of and parallel to the present tunnel. The east portal can be reached by a short walk up from Afton, while the west portal is accessible (on foot) from the CSX bridge over U.S. 250 on the west slope.

Troy & Greenfield's Hoosac Tunnel (1876) at North Adams, Massachusetts

Northern Massachusetts had long sought a route of its own to the west, and completion of the Western Railroad to the south in 1841 had only intensified that desire. Blocking the way was the 2566-foot Hoosac Mountain, which stood astride the natural route across the Berkshires between the Connecticut and Hudson river watersheds.

Titled "Work at the Heading," this drawing from *Scribner's Monthly* for December 1870 showed the work of rock drilling and mucking in progress in the Hoosac Tunnel. The Burleigh compressed air drills shown at work here helped make completion of the tunnel possible.—Author's collection.

An early construction view shows timber arch centering at the tunnel's west portal.—Smithsonian Institution (Neg. 90-16526).

By 1845 the Fitchburg Railroad had completed a northern route from Boston to Fitchburg, and a few years later the Troy & Greenfield was chartered to build west to the state line, right through Hoosac Mountain. The man behind both projects was Fitchburg manufacturer Alvah Crocker, whose determined efforts would eventually make the Hoosac Tunnel a reality.

The alignment selected for the tunnel took it 25,081 feet (4.75 miles) through the limestone, gneiss, and slate of the broad mountain. The line rose on grades of 0.5 percent from each portal to the center, climbing up from the valley of the Deerfield River to the east and the Hoosac River to the west. Designed to carry two tracks, the tunnel was 20 feet high and 24 feet wide.

Work began early in 1851 by manual drilling and black powder blasting, and it soon became apparent that Hoosac Mountain was not to be easily or quickly conquered. In an effort to improve progress, chief engineer A. F. Edwards brought in a massive steam-powered drilling machine. Edwards estimated that the tunnel could now be completed in just two years, but the machine traveled only about 10 to 12 feet before quitting for good.

Little was accomplished until 1856, when former Pennsylvania Railroad chief engineer Herman Haupt took over as both chief engineer and contractor for the project. Haupt struggled ahead with conventional tunneling and tried several more tunneling and drilling machines, none of which proved successful, before finally giving up the work in 1861.

The State of Massachusetts took over from the now-bankrupt railroad in 1862. Thomas A. Doane came in as the tunnel's third chief engineer and introduced the new methods that would finally complete the work. In 1866 Doane tried out a new compressed air drill invented by Charles Burleigh—the first practical mechanical rock drill in America—and these were soon being used in batteries of four to six mounted on a carriage designed by Doane. At the same time, Doane began using the newly developed nitroglycerin in place of black powder, and the combination of the Burleigh drills and the better explosive nearly tripled the rate of progress.

Doane left the project in 1867. The following year a contract to complete the work was awarded to W. and F.

Thomas Doane (1821–1897) was born at Orleans, Massachusetts. After completing studies at academies on Cape Cod and in Andover, he spent three years learning civil engineering in the office of Samuel M. Felton in Charlestown, Massachusetts.

Doane began his career as an engineer for railroads in Vermont and New Hampshire. He then returned to Charlestown in 1849 to set up his own engineering and surveying practice, working chiefly for railroads. His experience at Hoosac Tunnel in the development of compressed air machinery made him one of the pioneers in the use of compressed air in this country. Tunneling expert Henry S. Drinker, in his 1878 book on tunneling, said that to Doane's "persistent energy, far-seeing sagacity, and his able management, we, in large measure, and, in fact, chiefly, owe the development and introduction into this country of the present advanced system of tunneling with machinery and high explosives."

Doane's later experience included location and construction work for both the Burlington and the Northern Pacific, before he returned east to continue his consulting practice at Charlestown for the remainder of his life.

A double-headed eastbound Boston & Maine freight train emerged from the tunnel's east portal not long before it was electrified in 1911.—Smithsonian Institution (Neg. 88-18475).

Shanly & Company of Montreal, and the firm soon had as many as 1000 men working in three shifts. Tunneling progressed from both portals, and in 1870 the firm completed a 1028-foot central shaft down to the tunnel grade, providing both improved ventilation and a way to drill from four tunnel faces.

There were problems with flooding. A fire in the central shaft in 1867 killed 13 men and brought the work there to a halt for more than a year. Altogether, 195 men would die in the construction of Hoosac Tunnel.

At last the tunnel was holed through on November 27, 1873, and the end was finally in sight. Thomas Doane, who had returned to the tunnel project as consulting engineer in 1873, ran the first train through the tunnel on February 9, 1875; and all remaining work was completed over the next year. After 25 years, the excavation of 2 million tons of rock, and the expenditure of some $14 million, Hoosac Tunnel was formally opened on July 4, 1876. The 4.75-mile tunnel was the longest yet completed in the Western Hemisphere and second in the world only to the 7.9-mile Mont Cenis Tunnel in the Alps, opened in 1871.

The Hoosac Tunnel route across Massachusetts later became part of the Boston & Maine. Smoke and gases were always a problem in the long tunnel, and the B&M electrified the double track through Hoosac in 1911. This lasted until 1946, when diesels and an improved ventilation system took over. In 1957 the tunnel was single-tracked to provide greater clearances for piggyback cars and other high and wide loads. In 1998 a tunneling contractor enlarged the historic tunnel to provide a 20-foot clearance height for double-stack container trains, as Guilford Transportation Industries, the current owner of the line, prepared for a new and greater role for Alvah Crocker's Hoosac Tunnel link to the west in a post-Conrail New England.

Hoosac Tunnel was listed on the National Register of Historic Places in 1973, and was named a National Historic Civil Engineering Landmark by the American Society of Civil Engineers in 1975.

Powered by Electro-Motive F3 diesel No. 4253 and two FT units, a Boston & Maine freight westbound to the railroad's Mechanicville, New York, yard exited Hoosac's west portal in June 1949. A few years later the tunnel was single-tracked to provide clearance for piggyback traffic, and in 1998 it was enlarged for the still greater clearances needed for double-stack trains.—William D. Middleton.

GETTING THERE

The west portal of Hoosac Tunnel is in North Adams, about 1 1/2 miles south of the center on Ashland Street (State Route 8A) and then left about 1/2 mile on an unmarked road. The east portal is reached by traveling 7 1/2 miles east from North Adams on State Route 2, left on Whitcomb Hill Road for 2 miles, and left again on River Road for 1 mile.

—Canadian National.

Joseph Hobson (1834–1917) was one of those accomplished nineteenth-century engineers who rose to eminence in the profession without benefit of a formal engineering education. He was born near Guelph, Ontario, and attended school there. Before he was 18, Hobson was apprenticed to a Toronto civil engineer and land surveyor. He passed the examination for land surveyor in 1855 and began a practice in both survey work and civil engineering. Over the next ten years he was engaged in railroad surveys in both Canada and the U.S. and worked on the construction of the Grand Trunk in Ontario.

In 1869, Hobson was appointed an assistant engineer for the Great Western Railway and began a career in railroad engineering that was to last the remainder of his life. By 1875 he had become the railroad's chief engineer. He joined the Grand Trunk when the two railroads were amalgamated in 1882, and it was this that brought him together with the challenge of the St. Clair River Tunnel, which was to represent the greatest achievement of his career. By all accounts a modest man, Hobson declined a knighthood offered by Queen Victoria at the completion of this great work.

By 1896 Hobson had become the Grand Trunk's chief engineer, and the notable works under his charge included the replacement of John Roebling's suspension bridge over the Niagara River with a steel arch span in 1897 and the rebuilding of Robert Stephenson's Victoria Bridge at Montreal during 1897-98. Both were accomplished without interruption to traffic. He retired in 1907 but remained a consulting engineer to the Grand Trunk for the rest of his life. He died at his home in Hamilton, Ontario, at the age of 84.

The Grand Trunk's St. Clair River Tunnel (1891) and Canadian National's New St. Clair River Tunnel (1995) at Sarnia, Ontario

The St. Clair River, which formed part of the waterway linking Lake Huron and Lake Erie, represented a sizable obstacle to overland transportation between Michigan and Ontario. The river was half a mile wide and carried heavy steamship traffic moving between the lakes.

The Grand Trunk made the crossing with car ferries. The railroad had reached Sarnia, Ontario, from Toronto in 1858, while the Chicago & Grand Trunk completed a

This imaginative drawing from *Scientific American* for August 9, 1890, showed the hydraulic tunneling shield at work under the St. Clair River.—Author's collection.

The eastbound *Atlantic Express* exited the St. Clair Tunnel's Sarnia portal with 0-10-0 No. 598 in 1893. Specially designed for use in the tunnel, these big locomotives burned anthracite coal to minimize smoke and gasses in the tunnel. They were not particularly successful, and the tunnel was electrified in 1908.—W. Ethelbert Henry Photo, National Archives of Canada (Neg. PA-28818).

line from Chicago to Port Huron, Michigan, on the opposite bank, by 1880. The ferries presented difficult operating problems as the railroad's traffic volume grew. They were expensive to operate and slow. During the winter, ice in the river often interrupted ferry operations, and traffic backed up in the yards on either side of the river.

Because of the heavy water traffic, neither a low level bridge nor a drawbridge seemed practical, while the flat land on both sides would have made a high-level bridge and its approaches a costly undertaking.

A tunnel seemed to be the best solution, and as early as 1882 the railroad began to study its feasibility. By 1886 the Grand Trunk was ready to proceed. The St. Clair Tunnel Company was incorporated to build the tunnel, with Joseph Hobson as its chief engineer. Hobson, who was then chief engineer of the Grand Trunk's Western Division, proved to be an inspired choice. While he had no previous tunneling experience, Hobson was an experienced engineer who was able to apply the fast developing technology of underwater tunneling to the St. Clair River situation to complete successfully what would be the world's first large-scale underwater tunnel.

A tunnel would have to be driven through a hard blue clay topped with quicksand, with as little as 16 feet of cover separating the tunnel excavation from the deep river above, and Hobson proceeded cautiously. He first attempted small test tunnels on both sides of the river, and soon ran into trouble. On the Michigan side of the river, the work had to be stopped because of the poor quality of the material. On the Ontario side, gas and water-bearing clay were encountered under the river, and the work stopped.

While the test tunnels had been unsuccessful, Hobson had learned much about the problems ahead. In 1888 the Canadian government agreed to subsidize the cost of the tunnel and guaranteed the interest on the tunnel company's bonds. With that, work proceeded in May 1888.

The tunnel would have to be drilled through a narrow layer of clay between bedrock and the bottom of the river. Although this was as little as 38 feet deep at some points, the engineers decided to try to locate the tunnel at least 10 feet above bedrock to avoid the gas that had been encountered in the test tunnel and to maintain 16 feet or more of clay above the 21-foot tunnel to prevent water intrusion. Under these difficult conditions, Hobson decided that the shield tunneling technology, together with the use of compressed air ground support and cast iron tunnel liners, was the best way to drill the tunnel. It was the first time that all three had been combined in a large-scale underwater tunneling project.

Hobson first tried to sink a large shaft down to the tunnel level on each side of the river, from which the tunneling shields could be advanced. This proved unsuccessful, and in January 1889 excavation was started for deep approach cuts down to the level of the tunnel on each side of the river. These were complete by July on the American side and two months later on the Canadian side. The huge shields, one on each side of the river, were then lowered down ramps into the cuts and the tunneling was begun. As it was finally planned, the tunnel would be 6050 feet long, with 2.0 percent grades descending into the tunnel from each end and with a section of about

1500 feet near the center of the tunnel on a 0.1 percent grade rising toward the Port Huron end.

Hobson's design for the shields was based upon that developed by Alfred Ely Beach for his experimental Broadway subway at New York. Constructed of inch-thick steel plates, each shield was 21 feet 7 inches in outside diameter, 16 feet long, and weighed 80 tons. The leading edge of each shield sloped back from the top to provide a projecting roof over the tunnel facing. A bulkhead or wall divided each shield into two compartments, while horizontal and vertical bulkheads ahead of this dividing wall provided platforms for the excavators to work on different levels of the tunnel heading. Doors in the lower part of the dividing bulkhead were used for access to the tunneling face and to remove the excavated material.

A ring of 24 hydraulic jacks was mounted around the outer perimeter of the shield. Shoving against the already completed cast iron tunnel lining, these advanced a shield 2 feet at a time through the clay of the river bottom with a total force of 3000 tons. A rotating erector arm behind the dividing bulkhead was used to install the cast iron tunnel liners. The tunnel rings were made up of 13 of these segments, each 18 inches wide, 2 inches thick, and weighing a thousand pounds, bolted together with a keypiece to form the 21-foot diameter ring.

Contractors were unwilling to bid on the risky project, so the Grand Trunk carried out the work with its own forces. As soon as the shields were in place, tunneling began on July 11, 1889, on the Port Huron side and on September 21 at Sarnia. Drilling continued around the clock on a three-shift basis. Because horses were unable to work under the high air pressures, mules were used to pull the carts of excavated clay out of the tunnel.

There were problems with quicksand and water leakage, but these were solved with high air pressures. Some of the men were afflicted with caisson disease; three died from the effects and others were crippled. Hobson was worried about encountering explosive gases. The tunnel was fitted with an elaborate ventilation system, and miner's safety lamps were used to detect any gas infiltration.

Despite the problems the work progressed rapidly. Each shield averaged about 10 feet a day, and on one record day a shield was advanced almost 28 feet. After little more than a year's tunneling, the two shields met under the river on August 30, 1890. Hobson had devised an elaborate surveying system to assure that the shields stayed on line, and the two met with a deviation of less than a quarter of an inch.

Remaining work was completed over the following year, and the opening of the tunnel was celebrated with an inaugural train and elaborate festivities on September 19, 1891. The railroad soon found that the tunnel would save it $50,000 a year in operating costs over those for the car ferries and would cut two hours off the travel time of trains using the route.

Smoke and gases were expected to be a serious problem in such a long tunnel, and the company ordered special locomotives that burned anthracite coal, which would produce less smoke. These proved far from satisfactory, however, and by 1908 the tunnel had been electrified. Electric operation continued until 1958, when an improved ventilation system was installed and diesel power began operating through the tunnel.

Joseph Hobson built well, and his pioneer tunnel might have continued moving traffic under the St. Clair River indefinitely, except for the demands of modern freight traffic. Hobson's 21-foot diameter tube just wasn't big enough for present-day high and wide loads or the double-stack container cars that have become commonplace for intermodal traffic. Canadian National found that it was having to move more and more traffic across the St. Clair the old fashioned way on car floats, with all the extra cost and delay that this meant. With as many as 80,000 cars a year unable to use the old tunnel, CN began the construction in 1993 of a new tunnel that would have an internal diameter of 27 feet 6 inches, enough for the largest loads. In the technology and equipment used for its construction, the new tunnel would be every bit as much a pioneer for our time as its predecessor had been a century before.

To drill its new tunnel, CN acquired an extraordinary tunnel-boring machine named Excalibore that was 31 feet 3 inches tall, nearly 272 feet long, and weighed 796 tons. Guided by lasers, a satellite, and computers, Excalibore was capable of drilling through the clay of the river bottom at an average rate of about 26 feet a day. An articulated cutting head permitted the machine to change its route. The cutting head had 24 hydraulic cylinders, each capable of exerting a force of 220 tons. The machine had 200 permanent spade-type teeth and could be equipped with 53 backloading ripper teeth and 53 backloading twin-disc cutters. The machine was moved forward by 30 hydraulic cylinders, each capable of a force of 220 tons. Excavated material was removed from the tunnel by a screw conveyer and two belt conveyers.

The tunnel lining was made up of reinforced concrete rings, each weighing about 44 tons. Each ring was assembled by a rotating erector arm on the tunnel machine from six precast segments, each 5 feet wide, 15 feet long, and 16 inches thick, and a key segment.

The new 6130-foot tunnel was built alongside the original tunnel about 90 feet to the north, with the tunnel crown at about the same level and the bottom of the new tube about 10 feet lower. The line descended into the tunnel on a 2.1 percent grade on the Port Huron side and a 2 percent grade on the Sarnia side.

The tunneling contractor began boring from the Sarnia side of the river in September 1993. All went well until the end of the year, when work had to be halted for close to six months while a shaft was drilled down to permit replacement of some faulty seals on the tunneling machine. Underway again in August 1994, Excalibore

With unit No. 1305 in the lead, a pair of Baldwin-Westinghouse electric locomotives leads a westbound passenger train out of the tunnel's Port Huron portal.—Library of Congress.

GETTING THERE

The St. Clair tunnel portals are located in Port Huron, Michigan, and Sarnia, Ontario. The now-closed portals of the original tunnel are about 90 feet south of the new tunnel.

The Baltimore & Ohio's Howard Street Tunnel (1895) at Baltimore, Maryland

In the late nineteenth century the lack of a direct all-rail route through Baltimore put the B&O at a competitive disadvantage, particularly after the rival Pennsylvania Railroad completed two tunnels in 1873 that gave it a direct line through the city. The B&O still had to ferry all of its through traffic a distance of three-quarters of a mile across the West Branch of the Patapsco River between Locust Point and Canton.

There were proposals for a tunnel under the harbor or an elevated railroad across downtown Baltimore to provide the needed connection. A better solution emerged in 1887, when the B&O proposed a belt line around the city that would pass through downtown Baltimore in a long tunnel. By late 1890 the project was underway.

The masonry tunnel was built with an arched section 27 feet wide and 22 feet high to accommodate a double-track line. Beginning at the B&O's Camden Station, it extended north under Howard Street through the heart of the city for a distance of 7339 feet to a point south of Mt. Royal Avenue, where a tunnel "bell mouth" opened into a wide cut at the new Mount Royal Station. Beyond the station, two 265-foot double-track tunnels continued under Mt. Royal Avenue to the north portal, 8350 feet from the Camden Station portal. Over its full length the tunnel climbed at a uniform 0.8 percent eastbound grade from Camden Station.

While open cut, or "cut and cover," tunneling would have been more economical, this seemed infeasible for

far exceeded its design capacity, setting a record production rate of 21 rings—104 feet—on December 1, 1994. Just a week later the machine broke through on the Port Huron side, within an eighth of an inch of its planned alignment.

CN's new St. Clair River Tunnel began handling trains on May 5, 1995. With a new capability that saw as much as 14 hours being cut from some Halifax-Chicago stack train schedules, the St. Clair Tunnel route was soon handling a gratifying increase in traffic.

Just hours before the inaugural train operated through the new tunnel, the Chicago-bound VIA Rail/Amtrak *International* made the final trip through Joseph Hobson's historic tunnel. The tube was then sealed, although its ornate stone portals will be preserved. This is only appropriate, for the tunnel was listed on the National Register of Historic Places in 1970 and was designated a National Historic Civil Engineering Landmark by the American Society of Civil Engineers in 1991.

These drawings from the December 19, 1891, issue of *Engineering News* show typical construction methods used for the Howard Street Tunnel.—Author's collection.

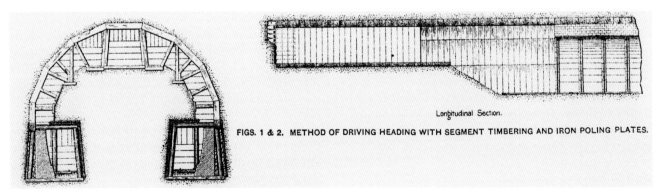

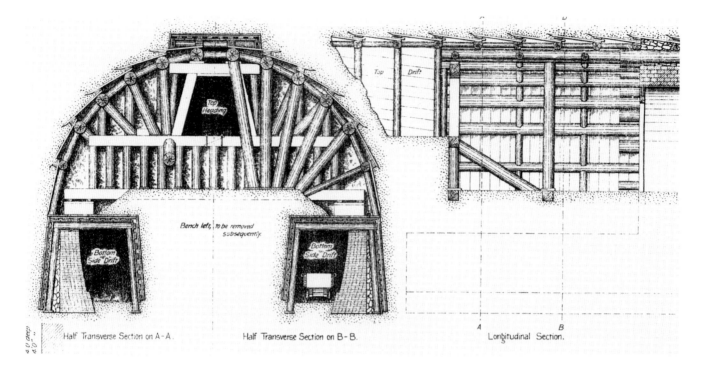

Half Transverse Section on A-A.   Half Transverse Section on B-B.   Longitudinal Section.

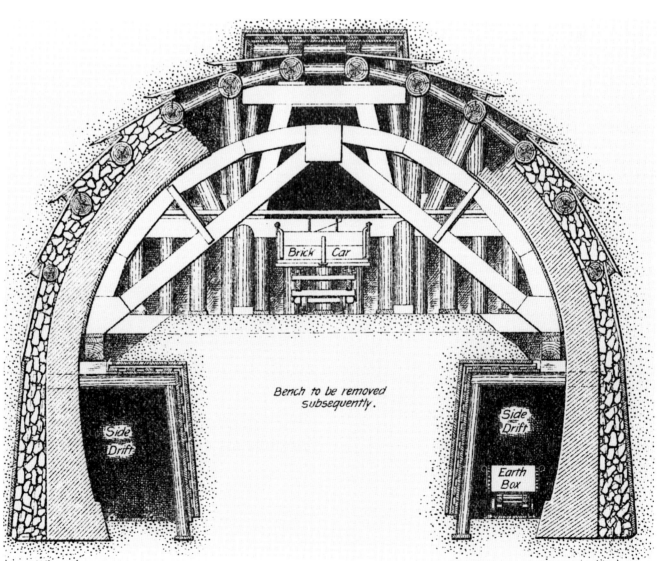

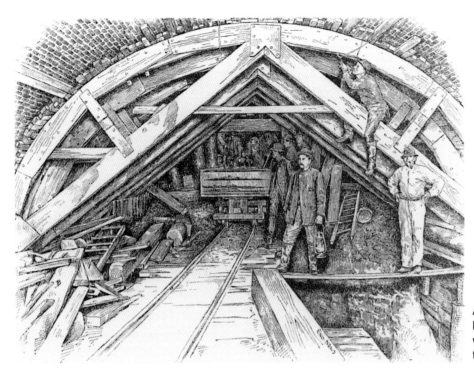

A drawing from *Engineering News* for December 19, 1891, shows workmen with arch centering in the interior of the tunnel.—Author's collection.

most of the tunnel. At some points, the crown of the tunnel would be as much as 70 feet below ground level, far too deep for easy open cut excavation, and the disruption of open cut work in the busy downtown street would have met with vigorous opposition. Instead, the B&O decided to drill almost the entire length of the tunnel. At the time it was completed, it would rank as the longest soft ground tunnel ever driven.

The tunnel was drilled through what was predominantly fine sand, intermingled with seams of gravel, clay and loams. Some of the clay was so hard it had to be blasted for removal. The tunnelers used the German method, in which bottom side drifts were driven first and then enlarged to the springing line of the roof arch. Driving these first helped to drain water from the soil of the upper section. Next, a center top heading was driven, and the haunch sections were taken out. The center core at the bottom was excavated only after the side walls and roof arch were completed.

Both the side drifts and top headings were strutted with heavy timbers until the permanent tunnel lining was completed. Side walls of brick masonry on a concrete foundation were built first. Timber or iron arch centers were then erected to support construction of the roof arch. This was usually laid up in sections of about 18 feet long and was made up of five rings of brick, although as many as eight rings were used in areas of unstable soil. Following excavation of the center core, an invert of concrete and brick was laid in the floor of the tunnel.

To permit tunneling to go forward from several headings, five access shafts equipped with hoists or elevators were sunk to the tunnel level. Except for one shaft located within Howard Street, all of these were sunk from vacant lots or the basements of homes or other buildings along the tunnel alignment.

About 1200 feet of tunnel near Camden Station that was too shallow for a drilled tunnel was constructed by the open cut method. Heavy traffic made it impractical to consider closing the entire street for this work. Instead, side trenches were first excavated and the masonry side walls built. Once these were in place, the center section was excavated and a timber structure erected between the side walls to provide a temporary street

Electrification made the long Howard Street Tunnel practical. This drawing from *Scientific American* for August 10, 1895, shows one of the B&O's big General Electric locomotives pulling a train through the tunnel.—Author's collection.

surface. The tunnel was then arched over and the street filled in and repaved as quickly as possible.

Tunneling work was started in September 1891. Great care was required to avoid disturbing traffic in Howard Street, which carried both a cable railway and a horse car line, as well as heavy wagon traffic. The foundations of adjacent buildings, some of which were eight to ten stories high, had to be protected against damage or settlement. Large quantities of water were encountered and pumps were installed to remove water from the excavation. Wells were sunk ahead of the tunnel heading and much of the water was pumped out before excavation began. At one point the water infiltration was so heavy that perforated pipes were driven into the surrounding ground and a cement grout pumped in to solidify the ground and block the flow. Settlement of the wet, loose material around the tunnel was a constant worry; and the tunnel contractor was frequently required to repair the street, building foundations, or water and sewer lines that had cracked or settled. At some points the street settled by as much as 12 to 18 inches. One building collapsed and had to be replaced, but little other major damage was done.

The tunnel construction attracted hundreds of interested spectators, reported an enthusiastic *Maryland Journal*. "The road cannot be but of immense benefit and convenience to the entire city of Baltimore," it concluded. "When finished the wonder will be why anyone opposed its construction."

Smoke and gasses from steam locomotives had promised to be a major problem in the long tunnel, and in 1892 the B&O made a daring decision to operate the tunnel with electric power. General Electric completed what would be the world's first electrification of main line steam railroad passenger and freight operation, and on May 1, 1895, the first train operated through the tunnel.

The tunnel has served the B&O and CSX well ever since. Diesels replaced electric power in 1952, and the tunnel has been converted to single-track operation to provide greater clearances for high and wide freight cars. In the early 1990s some tricky reinforcing work was required when Baltimore's new Central Light Rail Line was built in Howard Street, right above the tunnel.

The Howard Street Tunnel was placed on the National Register of Historic Places in 1973.

GETTING THERE

A relocated south portal of the tunnel is just south of the Camden Yard parking lot and is best seen from the Hamburg Street bridge. The north portal at the old Mt. Royal Station, now the Maryland Art Institute, lies between Preston and Dolphin streets and can be seen from the Dolphin Street bridge.

The Canadian Pacific's Spiral Tunnels (1909) in Kicking Horse Pass, British Columbia

The Canadian Pacific's choice of a southerly crossing of the Rockies through Kicking Horse Pass, together with a shortage of money and the need to complete the transcontinental line as rapidly as possible, had left the railroad with some difficult operating conditions in the pass.

The builders ran into problems almost immediately when they began to build the line down from the summit through the Upper Canyon of the Kicking Horse in 1884. Major A. B. Rogers had surveyed a route for the line that maintained the required 2.2 percent maximum grade, but it would have placed the track high on exposed slopes subject to avalanches, and it would have required a hard rock tunnel 1400 feet long through Mount Stephen. The line would have cost an extra $2 million to build, and it would have taken a year longer.

Instead, the CP built a 4 1/2-mile "temporary" line that descended from Wapta Lake, near the summit, to the base of Mount Stephen on a 4.4 percent grade. Safety tracks were installed at three locations to prevent runaways.

The "Big Hill," as it became known, was an operating nightmare from the beginning. The very first construction train that operated down the hill ran away, falling into the Kicking Horse River and killing three men. There were many more accidents to follow. One writer, Stuart Cumberland in his *The Queen's Highway*,

This view of an eastbound passenger train of the 1890s conveys a sense of the operating difficulties faced by Canadian Pacific on the original line through Kicking Horse Pass. Three locomotives were required to move eight cars up the 4.4 percent grade of the Big Hill. The train is near the crest at the No. 1 safety track. The runaway spur track is behind the rear pusher engine.—British Columbia Archives (Photo No. B-00230).

A construction photograph dating to about 1908 shows a crew at work at the portal to Tunnel No. 2, the lower spiral tunnel.—Glenbow Archives (No. NA-4598-7).

From Ocean to Ocean, 1887, commented, "For downright rugged awfulness, there is nothing on the whole of the Canadian Pacific Railway to equal the Kicking Horse Pass."

As CP traffic grew, the Big Hill became an increasingly serious bottleneck. Freight trains were limited to a maximum speed of 6 mph, passenger trains to 8 mph, on the hill. In 1908, it was reported that even four of the huge Baldwin 2-8-0 Consolidations then working on the Hill could handle a train of only 710 tons up the grade, and it took an hour to do it. It was time to bring the "temporary" arrangement in Kicking Horse Pass to an end.

The railroad considered an entirely new line through one of the other passes. There was a proposal for electrification of the line and another for a 10-mile tunnel through the pass. The best answer seemed to be to leave the line right where it was and to find a way to lengthen the line in order to reduce the gradient.

The solution, developed by CP assistant chief engineer J. E. Schwitzer, was patterned after a similar project on Switzerland's St. Gotthard Railway. Schwitzer's scheme halved the grade to 2.2 percent compensated by doubling the length of the Hill to about 8.2 miles. This was achieved in the narrow valley by building a new line that twice doubled back on itself, reversing direction in two spiral tunnels in which the line circled around over itself, one tunnel on each side of the canyon. In effect, the line was much like a double switchback, but with a great tunneled loop at each change of direction instead of the usual reversing stations.

Beginning at Wapta Lake west of the summit, the new line descended along the south side of the valley of the Kicking Horse for 3 miles, benched in the solid rock of the mountainside, then turned south into Cathedral Mountain in Tunnel No. 1, the upper spiral tunnel. Within the 3206-foot bore the line circled around through 234 degrees on a 10-degree curve (a 573-foot radius), exiting the mountain on a northwest heading through a portal 48 feet below the line above. The line then turned to the northeast to head back up the valley, still descending on a 2.2 percent grade.

A little more than a mile back up the valley the line curved back to the northwest to cross over the Kicking Horse and then curved around to the right through Mount Ogden in Tunnel No. 2, the lower spiral tunnel. Within the 2890-foot tunnel the line circled through 232 degrees of curvature on a 10-degree curve, finally exiting the mountain through a portal 45 feet below the line above. In both spiral tunnels the grade was reduced to 1.6 percent compensated.

Below the second spiral tunnel the line then crossed over the Kicking Horse again to continue down the south side of the valley, passing through one more new tunnel before rejoining the original line just above the Mount Stephen Tunnel. In addition to more than 8 miles of new line and the three tunnels, the work included four new crossings of the Kicking Horse River.

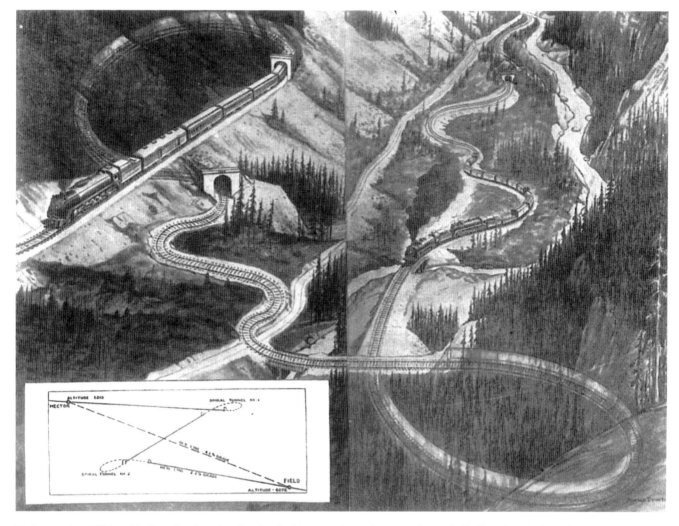

Dating to the 1930s, this imaginative drawing by A. Proctor shows how engineer J. E. Schwitzer's ingenious spiral tunnels conquered the rugged grades of Kicking Horse Pass. The view is to the southwest down the valley, showing the upper spiral tunnel on the left and the lower spiral tunnel to the right. Running diagonally through the illustration is the grade of the Kicking Horse Trail, the first improved road through the pass.—British Columbia Archives and Records Service (Neg. 23693 A-8713).

High above the valley of the Kicking Horse, an eastbound passenger train had the summit almost in sight just after leaving the upper portal of the upper spiral tunnel about 1940. —Nicholas Morant, Canadian Pacific Archives (Neg. M.2739).

With 2-10-2 No. 5810 and 2-10-4 No. 5930 in the lead, the first section of train 8 east to Hector crossed over the lower portal of the lower spiral tunnel at Yoho, British Columbia, on June 26, 1949.—Walter H. Thrall, Donald Duke Collection.

A Vancouver contractor began work in September 1907. Close to a thousand men were engaged on the project, and construction camps were set up at five locations in the pass. Some 400,000 cubic yards of rock were excavated to build the new line and tunnels, and the contractor used about 75 carloads of dynamite and enormous quantities of compressed air to do it.

Drilling of the tunnels began in January 1908, and both had been drilled through by June 1909. They were drilled through a crystallized limestone with Little Giant compressed-air rock drills, shot with dynamite, and then mucked out with Marion steam shovels fitted with special short booms and dipper arms operated by compressed air. Compressed air and electric lighting plants were set up at each tunnel, and the work continued day and night from both ends until the tunnels were complete. In places the limestone was badly crushed, and about a quarter of each of the spiral tunnels had to be timbered.

The new line opened to traffic on August 25, 1909. While it had taken four big 2-8-0s to move 710 tons over the old line, two of the same engines could now haul a train of 1490 tons up the grade. The Big Hill in Kicking Horse Pass had been conquered at last.

Schwitzer, who later became the CP's chief engineer, had planned well, and the ingenious tunnels have served the railroad effectively ever since. Subsequent modifications included the installation of a concrete and sprayed-on Gunite tunnel lining and, in 1992, the enlargement of the tunnel crowns to clear double-stack container cars.

GETTING THERE

The Trans-Canada Highway generally follows the Canadian Pacific through Kicking Horse Pass in Yoho National Park west of Lake Louise. Just west of the summit there is a public overlook to the spiral tunnels and a historical display on the highway. Graeme Pole's book (see Bibliography) provides a detailed guide to good viewpoints all through the pass.

The Pennsylvania Railroad's Hudson River Tunnels (1910) at New York, New York

At the turn of the century, advances in the technologies of underwater tunneling and electrification had finally offered the Pennsylvania Railroad a practical solution to its long search for access across the Hudson River to Manhattan. The railroad's massive New York tunnel and terminal project included twin tunnels under the Hudson that would take trains to a new Pennsylvania Station in Manhattan, while four tubes under the East River would link the station with the Long Island Rail Road and a new storage and servicing yard in Queens. All would be powered by electric traction.

The Hudson River tunnels offered some special challenges. These would extend an overall distance of 2.76 miles from a New Jersey portal on the west side of Bergen Hill to a point under 32nd Street at Ninth Avenue in Manhattan, while the tubes under the river would each be 6575 feet long. British tunneling engineer Charles M. Jacobs was appointed chief engineer for the work.

The tubes would have to be deep enough to clear the dredging plane established by the War Department, 40 feet below mean low water; they would also have to pass below the Manhattan shore line at a sufficient depth to avoid any damage to existing piers and bulkheads; and they would have to be far enough below the river bottom to avoid "blow outs" during compressed air tunneling.

To reach a sufficient depth, the tunnels descended on a 1.3 percent grade from the Bergen Hill portal to a low

A view of the rear of a shield in the south tube, advancing from Manhattan to a junction with a shield being driven from the New Jersey side, shows the rotating erector arm used to place the cast iron tunnel segments.—Smithsonian Institution (Neg. 94-4982).

point under the river 97 feet below mean high water and then climbed on grades of up to 1.93 percent toward Manhattan. This alignment placed the tubes in a fluid silt composed chiefly of clay, sand, and water; and the engineers selected shield-driven, compressed air tunneling as the most suitable for the work.

Jacobs designed 193-ton cylindrical shields for the project that were 23 $1/2$ feet in diameter and 17 feet long, with a 2-inch thick cast steel cutting edge. These were advanced by 24 hydraulic rams, each capable of exerting a force of 3400 tons. Doors in the front of the shield could be opened to admit material from the tunneling face.

A rotating hydraulic erector was used to place the cast iron segments that formed each tunnel ring behind the shield. Each ring was made up of eleven bolted segments and a closing "key" segment. Cast steel segments were used at points of unusual stress. After the tunneling was complete, this cast iron shell was lined with 2 feet of concrete.

The engineers were concerned about stability of the tubes under heavy railroad loads in the soft silt bed of the Hudson and considered several schemes for supporting the tunnels with piers or piling from a solid foundation in the river bed. In the end, the railroad decided that the extra support wasn't needed, and to this day the two tunnels float up and down a fraction of an inch with every change of the tide.

Construction of access shafts began in mid-1903, and the first shield began its advance from Manhattan in May 1905. Four shields were used, advancing from each end of the twin tubes to meet at the center of the river. More than a few problems were encountered. On the Manhattan side, the rock covering over the tunnel was

A May 9, 1907, construction photograph shows tunnel workers caulking joints in the cast iron tunnel segments of the south tube under the Hudson. The two men at the left are tightening the bolts that hold the segments together, while the four men at the right are installing "grummets" of red lead-soaked yarn behind washers to make the bolted joints watertight.—Smithsonian Institution (Neg. 94-4987).

Railroads below Ground

A view of the completed westbound tube under the Hudson approaching the Weehawken shaft. Concrete benches on either side of the track were designed to confine a train to the center of the track in case of a derailment, and provided walkways for emergency exit from the tunnel, on the left, and for signal maintainers, on the right.—Museum of the City of New York (Neg. 5361).

After passing under the Hudson, the twin tunnels were drilled through more than a mile of solid traprock under Bergen Hill before coming to the surface at Hackensack. Headed by DD1 electric locomotive No. 26, a 4 p.m. Philadelphia Express emerged through the Hackensack portal on June 22, 1913.
—Charles B. Chaney, Smithsonian Institution (Neg. 2140).

pierced and water flooded in, causing the ground above to subside under a New York Central freight yard. As the tubes pushed out under the Hudson from Manhattan, there were several blowouts of compressed air, allowing water to enter the tunnel from the river. During one of these the escaping air created a 20-foot geyser in the river, and water rose to a depth of 4 feet in the tunnel before it could be stopped.

Jacobs had planned to force the shields through the soft material without taking any material into the tunnel, but the shields tended to rise; and it proved impossible to keep the tunnel on the correct grade. Jacobs experimented with opening the doors to admit some of the excavated material, and it was found that the tubes could be steered up or down to maintain the proper grade simply by admitting more, or less, silt into the tunnel.

The work proceeded rapidly as the tunnelers gained experience. With crews at work on three shifts, the average rate of progress on each tube was about 18 feet per day. The two shields for the north tube met under the river in September 1906, a full year ahead of schedule; the south tube was joined a month later. Waterproofing and lining of the tubes was complete by June 1909, and the installation of track, electrification, and signaling followed.

The entire tunnel and terminal project was complete the following year, and on September 27, 1910, the Pennsylvania began operating its trains through the Hudson River tubes to Manhattan's Pennsylvania Station.

In 88 years of uninterrupted operation since then, Charles Jacobs's Hudson River tunnels have safely car-

—*Library of Congress.*

Charles Mattathias Jacobs (1850–1919) was one of the foremost tunneling engineers of his time. He was born at Hull, Yorkshire, England, and was tutored at home before beginning an engineering apprenticeship with a ship and engine building firm. At its completion in 1871, he supervised bridge erection in China before returning to England to establish a consulting engineering practice.

In 1889 Jacobs was invited to the U.S. by Austin Corbin, president of the Long Island and Philadelphia & Reading railroads, to advise the Reading on the manufacture of coal briquettes. His engineering skills were soon being used on other projects, among them a proposed LIRR tunnel under the East and Hudson rivers. Nothing came of that tunnel, but in 1892 Jacobs became chief engineer for the construction of a gas line tunnel under the East River that was the first tunnel successfully driven under either the East or Hudson rivers. Several years later he was appointed chief engineer for the construction of the Hudson & Manhattan tubes between New York and New Jersey. His work on these projects earned Jacobs the reputation that brought him the appointment as the Pennsylvania's chief engineer for the Hudson River tunnels. Still other achievements in his distinguished engineering career included the construction of major tunnels in France and Mexico, a 270-mile petroleum pipeline in India, and the Hales Bar Lock and Dam on the Tennessee River.

ried millions of travelers to and from America's largest city. Today, the role of this enduring work is undiminished, for the tunnels anchor the north end of Amtrak's busiest corridor and support a flood of New Jersey commuters that grows more intense every year.

GETTING THERE

The Bergen Hill tunnel portal in New Jersey can be seen from Tonnelle Avenue (U.S. 1 and U.S. 9) in North Bergen. The Manhattan portal is hidden under buildings, although a short open section of the Penn Station yard is visible between 31st and 33rd streets and west of Ninth Avenue.

The Michigan Central's Detroit River Tunnel (1910) at Detroit, Michigan

At the turn of the century the Michigan Central faced a difficult operating problem at its crossing of the Detroit River between Detroit and Windsor, Ontario. The need to ferry cars across the river was becoming an increasing handicap as traffic grew on the railroad's route across southern Ontario. The operation was costly and time-consuming at the best of times, but ice conditions during the winter months often delayed traffic indefinitely.

The railroad had been seeking a better way across the river for a long time. A tunnel crossing was begun in 1872, but the work was given up a year later. Over the next 30 years there were several efforts to plan a bridge project, but requirements for navigational clearances in the busy waterway made any bridge a difficult and costly project.

By 1904 the rapid development of electric traction had led to a new interest in a tunnel, and the railroad appointed a committee of engineers to report on its feasibility. Their report recommended the development of a tunnel, and by the following year the Michigan Central had begun surveys along the proposed tunnel line and had organized the subsidiary Detroit River Tunnel Company to carry out the work. William J. Wilgus, chief engineer of the New York Central, was appointed chairman of an advisory board of engineers to guide the work. Wilson S. Kinnear, the railroad's assistant general manager, was named chief engineer of the tunnel company.

Proposals for the work were solicited from contractors early in 1906. The railroad's plans for open cut and tunneled approaches on both sides of the river were based upon established construction methods, but an innovative new procedure was proposed for the portion of the tunnel under the river. Contractors were given four different alternatives for building this underwater section, or were allowed to come up with modifications or entirely new schemes of their own.

One alternative was conventional shield-driven construction, but the other three were all variations of a sunken tube, or "trench and tube," concept developed by Wilgus. All of these required the dredging of a trench in the bottom of the river, followed by the positioning of either prefabricated concrete tunnel sections or some type of tunnel section forms, which then would be filled with concrete. Never before had a major tunnel been built in this manner.

The method used by the successful contractor was still another variation of the Wilgus sunken tube scheme. For the 2668-foot-long underwater section, a trench first was dredged in the blue clay of the river bottom along the tunnel alignment. Large twin-tube steel tunnel sections were then floated into position over the trench and

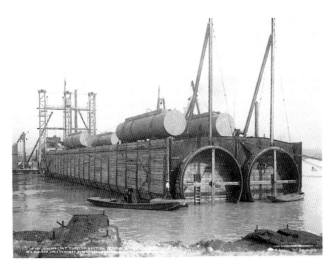

A Detroit Publishing Company photographer was on hand to record the sinking of the final tubular section for the Detroit River Tunnel in 1909. Here, the huge section is floated into position. The two tubes are bulkheaded with timber to make the structure buoyant, while the timber sides will act as forms when concrete is placed around the submerged structure. The four air tanks atop the section will be used to help control the descent during sinking.—Library of Congress (Neg. LC-D4-71383).

A crowd gathered on the shore to watch the descent of the final section. Beyond the partially submerged section is a large work barge with a tower structure that will be used to place concrete around the twin tube structure once it is in position on the river bottom. Concrete will be mixed on the barge and placed under water through tremie pipes—Library of Congress (Neg. LC-D4-71384).

sunk into place. These were then used as a form for exterior concrete, which was placed from a barge positioned above the tunnel section.

The steel tunnel sections were constructed at a shipyard and towed to the tunnel site. The twin tubes for these sections were each 23 feet 4 inches in diameter and were riveted together from 3/8-inch steel plate with a system of steel angle stiffeners. Steel diaphragm sections attached to the exterior of the tunnel sections at intervals

—Library of Congress.

William John Wilgus (1865–1949) was a largely self-made engineer. Born at Buffalo, New York, he graduated from high school there and then studied for two years under a Buffalo civil engineer. Wilgus began his career in railroad civil engineering at the age of 20, first with the Minnesota & Northwestern and then with other midwestern lines. He joined the New York Central & Hudson River in 1893 and rose rapidly through the engineering ranks. By 1899 Wilgus was its chief engineer. Over the next several years he planned the electrification of the Central's lines entering New York City and the new Grand Central Terminal that would represent his greatest achievements.

Wilgus left the New York Central several years later to practice as a consulting engineer. During World War I he served as director of military railways in the American Expeditionary Force. Shortly after the war he had an early role in planning the Holland Tunnel at New York.

His many awards included the Telford Gold Medal of the Institute of Civil Engineers of Great Britain. He was an honorary member of both the American Institute of Architects and the American Society of Civil Engineers and won both the latter's Rowland Prize and its Wellington Prize. He died at Claremont, New Hampshire, at the age of 83.

supported heavy timber sheathing that formed a sort of outer "hull" for each section. The overall size of each section was about 56 feet wide and 31 feet deep. Except for shorter drainage sump and closing sections, each of the tunnel sections was 262 feet 6 inches long and weighed about 600 tons. Open at the top and bottom, these structures acted as outer forms for placing the underwater concrete.

A contract was awarded on August 1, 1906, and work began two months later. The water depth at the tunnel site was as much as 50 feet, and a 48-foot-wide trench for the tunnel sections was excavated anywhere from 30 to 50 feet into the river bed with a large clamshell bucket crane mounted on a barge. Next, a grillage of steel I-beams was placed on the bottom and concreted into position as a temporary support for each end of the tunnel sections. This was done from a large work barge that also carried the mixing plants for the concrete that would be placed in the underwater tunnel

Two Michigan Central passenger trains are shown at the Detroit portal of the tunnel in this classic glass plate Detroit Publishing Company view. R-1 class B+B tunnel motor No. 8470 emerges from the tunnel with a westbound train, while on the upper level K-80 class 4-6-2 Pacific No. 7504 parallels the electric into the railroad's new Detroit passenger station.—Library of Congress (Neg. LC-D4-72270).

sections. The barge was fitted with two steel "spuds" that could be extended to the bottom to hold it firmly in position.

Sinking of the tunnel sections began from the Detroit side on October 1, 1907. The ends of the tubes were bulkheaded with timber, and four air tanks were attached to each section to help float it and guide it into position. Air was released from the tubes to sink each section into place gradually. Once each tube was in position, concrete was placed around it through "tremie" tubes from the concrete plants. The tunnel sections were connected by means of watertight joints with rubber gaskets. After the entire tunnel had been placed, the remainder of the trench on either side was backfilled with gravel and clay. The tubes were then pumped out and a concrete tunnel lining was installed.

Tunneling work was largely completed during 1909, and the installation of track and the electrification system was complete the following year. The first train passed through the tunnel on July 26, 1910, and regular freight and passenger operation began over the next several months.

For the Michigan Central, the tunnel scheme developed by William Wilgus was a splendid solution to its need for a Detroit River crossing that saved the railroad an estimated $2 million over what a more conventional shield-driven tunnel would have cost. Electric operation continued until all operations were converted to diesel power in 1953. The tunnel still serves Conrail well today, and the innovative technology that its builders pioneered has become one of the most commonly used methods of underwater tunneling.

GETTING THERE

The Detroit portal of the tunnel is a short distance east of the old Michigan Central station on 15th Street, while the Windsor portal is in the west end of the city between Ouellette Avenue and Huron Church Road.

The Denver & Salt Lake's Moffat Tunnel (1928) west of Denver, Colorado

There were no easy routes for a railroad across the Rocky Mountains of Colorado. Seeking easier grades, the Union Pacific built far to the north in Wyoming. The Santa Fe went to the south for the same reason. Even when Colorado finally got its own railroad across the Rockies, the Denver & Rio Grande Western went by way of a roundabout route to the south of Denver through Pueblo and over the highest crossing of the Continental Divide on any North American transcontinental at Tennessee Pass.

A direct rail route west across the Rockies was long a dream of Denver's civic leaders. No one wanted it more than David H. Moffat. The young businessman became interested in development of a railroad west from Denver soon after the UP had bypassed Denver in 1866, but not until 1902 was he able to begin construction of his Denver, Northwestern & Pacific.

It was an extraordinary mountain railroad. The line went over Rollins Pass west of Denver at an elevation of 11,680 feet that made it the highest standard gauge line

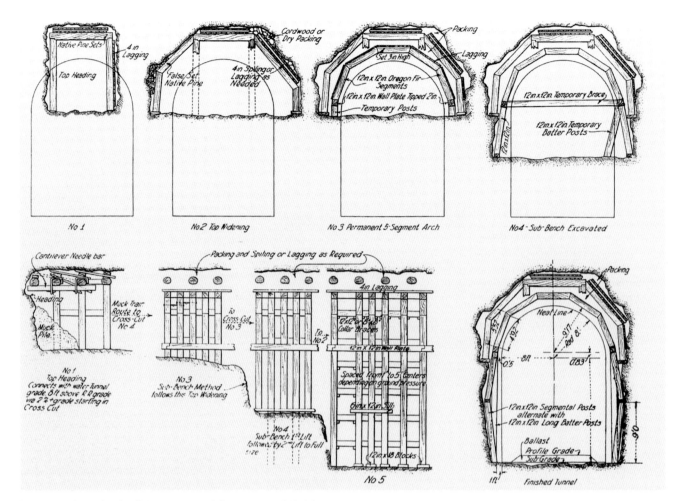

A drawing from the April 1925 issue of *Compressed Air Magazine* shows the typical stages of excavation and timbering in hard rock for the Moffat Tunnel, beginning with the driving of a top heading, followed by widening, construction of permanent timber arch sections, and excavation of the bottom sub-bench.—Library of Congress.

in North America. On the lower slopes of the front range the line snaked through a rugged terrain that required almost unending trestles, tunnels and sharp curves to maintain the desired 2 percent grade. Moffat planned to build a tunnel near the top of the pass later, so the 20 miles of line that climbed through the last 2000 feet to the summit were considered temporary and were laid on a 4 percent grade.

It was an operating man's nightmare. Snow could be expected any time from September to June, and it frequently filled some of the cuts to a depth of 40 feet. A long snowshed was built at the summit of the pass, and Moffat fought the snow and the grades with rotary snow plows, helper engines, and big Mallet locomotives; but it was always a difficult and costly operation. During severe winter weather the line was sometimes blocked for a month or more at a time.

The trains had begun running over Rollins Pass in 1904, but Moffat was never able to build his tunnel under the Divide or complete the line to Salt Lake City. He finally ran out of money in 1911, shortly before his death. Reorganized as the Denver & Salt Lake, his railroad made it to Craig, Colorado, and quit.

Denver still wanted its direct line to the west, and supporters of the tunnel continued to press for it. A Moffat Tunnel Bill finally passed in the Colorado legislature in 1922, establishing a tunnel improvement district with the authority to levy taxes and issue bonds to finance the project. An initial bond issue had been sold by mid-1923, and construction could begin.

The final alignment provided for a 32,799-foot (6.21-mile) tunnel under James Peak that would ascend on a 0.3 percent grade from the east portal to a maximum elevation of 9241 feet at mid-tunnel and then descend on a 0.9 percent grade to the west portal. The single-track tunnel would have a finished cross section 16 feet wide and 24 feet high. It would give the D&SL a new route over the Divide that was 23 miles shorter than the old line and had a maximum gradient of only 2 percent. It would reduce the maximum elevation of the crossing by 2406 feet, eliminate 11,000 degrees of curvature, and reduce the maximum curvature from 16 degrees to 9 degrees.

A contract was awarded in September 1923, and drilling began from the west portal before the end of the month and from the east portal during October. More

Moffat Tunnel workers are shown completing a heading through a hard rock section of the tunnel. An electric mucking machine has scooped up the loose rock after blasting and a conveyor is loading the material aboard a train of dump cars pulled by a battery-powered locomotive.—Smithsonian Institution (Neg. 97-589).

than 600 men were expected to be engaged in the work, and construction camps had been set up at both portals even before the contract was signed. There were a number of families living with the men at the camps, and these were self-sufficient communities, complete with schools, hospitals, housing, and recreational facilities. Work continued around the clock on a three-shift basis, and meals were available 24 hours a day in the camp dining halls.

The contractor drilled a smaller pilot tunnel in advance of the main tunnel on a parallel heading 75 feet to the south. Cross cuts were made to the main tunnel line at 1500-foot intervals to provide access to additional drilling faces, and the cross cuts and pilot tunnel were used to supply materials; to run electric power, water and compressed air lines; and to remove excavated materials from the tunneling faces. After the tunnel was completed, the pilot tunnel was used as an aqueduct to supply water to Denver from the west slope.

In soft material a top heading was driven first and then widened, followed by excavation of the lower bench. In hard rock a center heading was excavated first, and the tunnel then enlarged to its full size by the ring shooting method, in which rings of holes were drilled at right angles to the tunnel centerline. The holes were then filled with dynamite and the ring segments blasted away.

Compressed air drills mounted on carriages drilled the rock faces, which were then shot with dynamite. Electric mucking machines scooped up the excavated material and conveyor belts transported it back to be loaded on rail cars, which were than hauled out to the portals through the cross cuts and the pioneer tunnel by battery and trolley locomotives. The use of the parallel tunnels made it possible for drilling and mucking crews to alternate between headings in the two tunnels, so that one crew never had to remain idle while the other did its work.

The work from the west portal proved much more difficult than expected. Instead of hard rock, soft and unstable material was encountered, requiring almost

This photograph shows a completed heading through a hard rock section of Moffat Tunnel. A 10-foot bench remains to be removed before the section will be complete.—Smithsonian Institution (Neg. 97-598).

continuous heavy timbering to prevent cave-ins. The worst accident of the entire project occurred in this heading on July 30, 1926, when a sudden rock fall of some 125 tons killed six men. At some points the pressure from the overlying rock was so great that even the heavy 12 inch by 12 inch fir timbers were deformed or damaged. Harder rock was finally encountered more than a mile into the mountain.

The tunnelers at the west end were having a difficult time enlarging the tunnel section to its full size in the loose and unstable material. George Lewis, the tunnel commission's chief engineer and general manager, solved the problem with an ingenious device called the Lewis traveling cantilever girder. This consisted of two 48-inch steel girders, 60 feet long and braced 6 feet apart, with cross arms that supported the timbering in the tunnel crown while the bench below it was excavated and permanent timber posts installed to support the timbering above. Once the bench excavation had reached the maximum 20-foot overhang of the cantilever girders, they were pulled ahead on rollers to the next section. *Railway Age* called this the device that "turned defeat into victory."

The tunnelers working from the east portal found much harder rock that required very little timbering, and the work advanced rapidly until some serious problems were encountered from water that entered from crater lakes above the tunnel line. In February 1926 so much water entered the tunnel from one of these that the men had to abandon their equipment and run for safety. For six days the water flowed in at a rate of 3100 gallons a minute, flooding the tunnel and depositing silt along the floor for a distance of 1200 feet. Pumps had begun to control the flood when a blizzard knocked out electric power, and the tunnel flooded again. Power was restored and the tunnel almost cleared, when another power failure allowed it to flood once more. The tunnel was finally cleared and the hardened mud excavated, but the drilling had been delayed for three months.

On February 12, 1927, crews from the east and west portals holed through the pilot tunnel, and six days later "holing-through" ceremonies were held when President Calvin Coolidge set off a final blast by telegraph. The main tunnel remained to be completed. Three more Lewis traveling cantilever girders were put in service to

Denver & Rio Grande Western 4-8-4 Northern No. 1703 headed into the east portal of Moffat Tunnel with a Denver-Craig local in August 1950. —David P. Morgan Memorial Library.

deal with bad rock conditions, and by June, 15 tunnel faces were being worked at one time. The main bore was finally holed through on July 7, 1927.

The entire tunnel was lined with timber, steel, concrete, or Gunite. The portals were finished and the track laid. A high capacity ventilation system was installed at the east portal. The longest tunnel in the Western Hemisphere was now complete, and the tunnelers had excavated some 750,000 cubic yards of rock and used two and a half million pounds of dynamite to do it.

The first freight train passed through the tunnel on February 24, 1928, and two days later special trains arrived from Denver with a party of dignitaries and 2500 excursionists for the official opening of the Moffat Tunnel. David Moffat had been dead for 17 years, but he had not been forgotten; his name was set in concrete on the tunnel portals. The great tunnel he had conceived was finally complete.

But Denver still didn't have the direct route to the Pacific Coast that David Moffat had set out to build, for the Denver & Salt Lake still went only to Craig. This was finally resolved in 1934, when the Rio Grande completed the Dotsero Cutoff, which linked the D&SL with the Rio Grande's main line to create a new direct route between Denver and Salt Lake City. The Moffat Tunnel Route was later merged into the Rio Grande, and it remains today a key transcontinental route for Union Pacific. The tunnel was designated a National Historic Civil Engineering Landmark by the American Society of Civil Engineers in 1973.

GETTING THERE

East Portal can be reached via State Route 72 northwest from Denver and a road which follows the old D&SL alignment west from Rollinsville. The west portal is at Winter Park on U.S. 40.

The Great Northern's Cascade Tunnel (1929) at Skykomish, Washington

Although the Great Northern had found a relatively easy route across the Rocky Mountains, getting across the Cascade Mountains of western Washington was a different story. John F. Stevens had located the best route available at Stevens Pass, at the heads of the Wenatchee and Skykomish rivers, but it was still a difficult crossing.

Stevens located a line through the pass that maintained the desired 2.2 percent maximum grade until he reached a point only a few miles from the summit. From here it would take a 2 1/2-mile tunnel to cross the Cascades at a reasonable grade. Drilling it would take several years, but Jim Hill wanted the GN running to Puget Sound by the end of 1892. The only way to do it was to take the line over the summit on a temporary line and complete the tunnel later.

The temporary crossing Stevens laid out challenged the skills of the best of operating men. Even with no fewer than eight switchbacks in the line, maximum grades of as much as 4 percent and some severe curves were required. Operating over the pass on the steep switchbacks proved difficult at the best of times, but it was even worse in the severe winter weather that was commonplace in the Cascades. Moist clouds off the Pacific hit the west slope of the mountains and dumped anywhere from 8 to 12 inches of snow an hour on the pass. Miles of line were covered by timber snowsheds. Snowplows and hundreds of men worked to clear drifts as much as 75 feet deep.

This was not the kind of railroad Jim Hill had in mind, and work began on a tunnel almost as soon as the line over the top was completed. When the 2.63-mile Cascade Tunnel was finished in December 1900, it reduced the summit elevation by 677 feet, cut the maximum grade for the crossing to 2.2 percent, and eliminated 2332 degrees of curvature. The GN's route over Stevens Pass was now 9 miles shorter, and the new line had cut about two hours from the running time of all trains.

While Cascade Tunnel had improved operating conditions in the pass, it hadn't eliminated all the problems. Snow remained a formidable winter adversary. Despite miles of snowsheds, heavy snowfalls and avalanches continued to plague the line on the western slope of the Cascades. The avalanche danger became tragically real on March 1, 1910. Early that morning a massive avalanche at Wellington carried two stranded trains into the canyon below. More than a hundred people were killed, making it the worst U.S. avalanche disaster ever.

The Great Northern was soon planning another tunnel, much longer and at a lower elevation, that could avoid the worst of the winter operating problems and provide greater operating economies over the Cascades.

Dating to 1908, this view shows the rugged terrain of the Great Northern's Cascade Mountain crossing between Scenic and the Martin Creek trestle that would be bypassed by construction of the Cascade Tunnel.—H. L. Broadbelt Collection.

Entire camp towns were set up to house the hundreds of workers that built Cascade Tunnel. This was the construction camp near Scenic.—Great Northern, David P. Morgan Memorial Library.

After several years of study a tunnel route was adopted in the summer of 1925.

A new 10-mile line, including the tunnel, would connect with the existing line at Berne, on the east slope, and at Scenic, on the west slope. Trains would be powered through the tunnel by a new 73-mile electrification that extended over the entire Cascade crossing. The tunnel itself would be a single-track, 41,152-foot (7.79-mile) bore under Jim Hill Mountain on a 1.565 percent descending grade to the west. The new line and tunnel would reduce the summit elevation of GN's Cascade crossing to 2881 feet, 501 feet lower than the old line; and it would eliminate 1940 degrees of curvature and shorten the line by 7.68 miles. The new route would also get the GN entirely out of the avalanche area on the west slope, and eliminate the need for 6 miles of snowshed.

To complete the tunnel in the shortest possible time, the railroad planned to drill from both portals, from a shaft sunk to the tunnel elevation from Mill Creek, 2.41 miles from the east portal, and from cross-cuts driven at intervals of about 1500 feet from a parallel pioneer tunnel that would be driven from the west portal to the Mill Creek shaft.

This pioneer tunnel was drilled 66 feet south of the main tunnel line, with cross-cuts to the main tunnel. It provided access to multiple drilling faces for the main tunnel, and was used for air, water and power lines, and as a haulage way for supplies and muck. This reduced interference with excavation in the main tunnel and left it clear so that concrete lining work could follow closely behind the drilling. It also proved useful for drainage from the tunnel after heavy water flows from rock seams were encountered.

The tunnel was drilled largely through granite. The center headings were drilled from a mounted drill carriage that moved forward to the drilling face on a track. Each cycle or "round" of drilling, blasting, and muck removal took a little under five hours and advanced the heading 8 feet. Once the center heading was complete, the main tunnel was enlarged to its full size of 18 feet wide by 26 feet high by radial drilling and blasting. Electric mucking machines were used for the pioneer and

Drillers and blasters removed almost a million cubic yards of rock and earth to build the 7.79-mile tunnel. Here, the tunnelers attack a rock face with a battery of drills.—Great Northern, David P. Morgan Memorial Library.

The last shovelful of muck was taken out of the tunnel at about 7 a.m. on December 8, 1928. Photographer Lee Pickett was on hand to record the event.—Library of Congress (Neg. LC-USZ62-112928).

center heading work, while air-operated Marion shovels were used for mucking of the main tunnel enlargement. Muck was removed in trains of narrow gauge cars pulled by electric locomotives.

At Mill Creek a central shaft was driven 622 feet below the valley to reach the tunnel grade. Electric hoists supplied materials and removed muck from the shaft, and large "sumps" were excavated below the tunnel grade to store muck awaiting removal and to contain any drainage water until it could be pumped out.

GN awarded a contract for the tunneling late in 1925, and the contractor wasted no time in getting started. Within three months, camps had been completed at each portal and at Mill Creek, and more than 600 men were at work. At the peak of work, the contractor would have nearly 1800 men at work. The camps were virtually complete towns, each with a kitchen and dining room, bunkhouses for the unmarried workers, houses for married men and their families, hospital, commissary store, school, and recreation hall.

Drilling began from both portals in December 1925. The pioneer tunnel was advanced first from the west portal. Soft ground in this initial section necessitated continuous timbering, delaying progress; and it was decided to drill an inclined shaft or adit from the valley of the Tye River, 2270 feet in from the west portal, to accelerate progress for the pioneer tunnel. Drilling for the Mill Creek shaft began in February 1926 and was completed to the tunnel grade five months later, when drilling commenced in both directions.

Work continued day and night, and the tunnel workers proceeded to set records for tunneling speed. Center headings from the east portal and Mill Creek were "holed through" on March 4, 1927, the tunneling crews having driven 9533 feet at an average of 22 feet per day. Five times during 1926 and 1927 the workers set new world records for tunnel advances in a 31-day period, culminating in August 1927 with a 1220-foot advance in 31 days by the east portal crew enlarging the main tunnel section.

The 5.4-mile pioneer tunnel between the west portal and Mill Creek was holed through on May 1, 1928, when President Calvin Coolidge closed an electric circuit at the White House to fire the final dynamite shot. The main tunnel was holed through between the west portal and Mill Creek the following October and had been excavated to its full size by late November. By this time almost a million cubic yards of rock and earth had been excavated, and it had taken nearly five million pounds of gelatin dynamite and three-quarters of a million blasting caps to do it. The entire tunnel was lined with concrete at least a foot thick to provide a finished tunnel 16 feet wide and 20 feet 10 inches high above the track level. This work was complete in December 1928, and track laying and the electrification were complete early in January.

In just three years' time the Great Northern and its contractor had completed what would stand as North America's longest tunnel for the next 60 years, and they had set new tunneling records in the process.

The railroad celebrated completion of the work with appropriate festivities on January 12, 1929. Hundreds of guests arrived aboard special trains; there was a coast-to-coast NBC live radio broadcast; and the passage of

The heavy snow of a Cascade Mountain winter covered the ground early in January 1929 when Lee Pickett recorded the first train at Cascade Tunnel's west portal at Scenic, Washington.—Library of Congress.

the first train through the tunnel was followed by a banquet in the construction camp dining hall at Scenic. Among the speakers that night, none loomed larger than 75-year-old John F. Stevens, who was there to see the final triumph of the Great Northern in its long battle with the Cascade crossing he had discovered almost 40 years before.

The new Cascade Tunnel and the other line improvements brought the railroad's Cascade Division up to the high standards that Jim Hill had envisioned when he set out to build the Great Northern. "We can now regard our railroad as complete in all its parts," commented GN Vice President L. C. Gilman at the dedication banquet.

Cascade Tunnel confirmed the Great Northern's status as the best and most direct route to and from the Pacific Northwest, and it has served the railroad and its successors well ever since that opening day in 1929. Electric operation over the Cascades ended in 1956, after the railroad completed a ventilation system for diesel motive power. The Stevens Pass tunnels and John Stevens's switchback system were designated a National Historic Civil Engineering Landmark by the American Society of Civil Engineers in 1993.

GETTING THERE

The west portal is at Scenic, about 10 miles east of Skykomish on U.S. 2, which continues across Stevens Pass to the east portal at Berne.

The Canadian Pacific's Mount Macdonald Tunnel (1989) in Rogers Pass, British Columbia

The Canadian Pacific has been doing battle with the Selkirk Mountains of British Columbia for more than a century. The first battle against this formidable barrier was waged by Major Albert B. Rogers, who surveyed the route across the Selkirks through the pass that now bears his name, and whose opening in 1885 brought the Canadian transcontinental to completion (see page 93).

CP took on Rogers Pass a second time in 1916, when the 5-mile Connaught Tunnel was driven through the base of Mount Macdonald to avoid the pass altogether. This dropped the summit elevation by more than 500 feet to an elevation of 3745 feet and shortened CP's transcontinental line by 4 ½ miles.

Connaught Tunnel helped, but by the 1960s continuing traffic growth had made still further improvement necessary. It was 1983, however, before the railroad was able to move ahead with plans for its third crossing of Rogers Pass. This was a plan to build a new 21-mile line with a 1 percent maximum grade for westbound trains, this being the direction of CP's heaviest grain and coal traffic. Eastbound traffic would continue to use the line through Connaught Tunnel, with grades up to 2.2 percent.

CP achieved this new low-grade line with a 48,304-foot (9.1-mile) bore through Mount Macdonald that is the longest rail tunnel in the Americas. The work also included a second tunnel over a mile in length, five bridges, a 4032-foot viaduct, and 11 miles of track at grade. It took more than a thousand men four and a half years to build it. The engineering effort was led by CP vice president John Fox, with design work by CP's own staff and consulting engineers Parsons Brinckerhoff Quade & Douglas, who designed the long tunnel.

Construction of the tunnel was a showcase of late twentieth century tunneling technology. Built on an alignment 6000 feet below the mountain's peak, it was bored through beds of hard and abrasive quartzite and phyllite, and other rock. A horseshoe-shaped tunnel was excavated 19 feet wide and 29 feet high, and then lined with a 10-inch thick concrete wall and crown, and a 13-inch floor slab.

Two contractors drove the long tunnel. At the west end a Canadian-Japanese joint venture drilled 4 miles of tunnel at a rate of up to 50 feet a day using a conventional drilling and blasting technique, with some interesting variations. A six-boom drill jumbo mounted on rails and advanced and retracted by hydraulic jacks was used to drill about 100 holes at a time 14 feet into the tunnel face. These were then blasted, the drill jumbo retracted, and a huge, two-clawed excavator known as the "lobster" moved in to muck out as much as 10 cubic yards of rock a minute. One track laid in the completed tunnel carried muck trains, while a second brought in concrete for the tunnel lining.

A joint venture of U.S. and Canadian firms bored 5 miles of tunnel at the east end. A huge, 22-foot 4-inch diameter tunnel boring machine drove a top heading, and the bench below this heading was then removed by drilling and blasting to complete the tunnel. At first, the hard quartzite was "eating a cutter a foot," according to the contractor's project manager, limiting the daily advance to about 50 feet. But once the machine got into softer material, it made as much as 174 feet a day. Laser surveying equipment kept the two tunnelers precisely on the required alignment, and when the two bores met they were off by only 2 inches vertically and 6 inches horizontally.

A third contractor sank an 1143-foot ventilation shaft, 28 feet in diameter, down to the tunnel midpoint for the ventilation system that cools the tunnel and purges it of diesel locomotive exhaust fumes. This elaborate system reduces the time required to clear the long tunnel to about 30 minutes after the passage of a train. Five 2250 h.p. electric fans are used, four at the top of the midpoint shaft and one at the east portal. As a train enters the normally westbound tunnel from the east portal, a gate at the midpoint is closed, and air is drawn past the train by the fan at the portal, while two of the midpoint fans force air into the tunnel. As the train reaches the midpoint, the gate is opened and the midpoint fans are shut down, allowing the east portal fan to pull air past the train from the west portal. Once a train is past the midpoint, the gate is closed again, and two midpoint fans draw air past the train from the west

While a tunnel-boring machine was used at the east end of the Mount Macdonald Tunnel, the Japanese-Canadian joint venture contractors for 4 miles of tunneling at the west end drilled up to 50 feet a day using conventional drill and blast techniques, as shown here.—Canadian Pacific.

Once rock drilling was completed, double tracks were installed for construction trains. One was used for muck removal, while the other was used to bring in concrete for tunnel lining.—Canadian Pacific.

The tunnelers broke through on North America's longest railroad tunnel on May 6, 1987.—Canadian Pacific, David P. Morgan Memorial Library.

Work was nearing completion on the bridge that would carry the CP's new Rogers Pass line across a stream just outside the 9.1-mile tunnel's west portal.—Canadian Pacific.

portal, while the other two force air through the east half of the tunnel to clear it of fumes.

The principal tunneling contracts were awarded in May 1984. The two tunnel sections were linked with a final blast just two and a half years later, and on December 12, 1988, the first train rolled westward through the tunnel. Formal dedication ceremonies followed on May 4, 1989.

In a time of growing traffic and competitiveness, Mount Macdonald Tunnel was indeed a worthwhile investment. CP's third crossing of the Selkirks had increased capacity by about 60 percent, from 15 trains a day to as many as 24, as well as substantially reducing its operating costs. The westbound helper station at Rogers was closed, and running times over the railroad's Field-Revelstoke district substantially reduced.

GETTING THERE

Rogers Pass lies about 40 miles east of Revelstoke on the Trans-Canada Highway, which generally parallels the CP's route through the Selkirks.

138   **Landmarks on the Iron Road**

Approaching the east portal of Mount Macdonald Tunnel on the CP's new line through Rogers Pass, a westbound freight train crossed the new bridge over Stoney Creek.—Canadian Pacific.

Four massive Hulett unloaders unload the Great Lakes Steamship Company's ore boat *J. Burton Ayers* at the Pennsylvania Railroad ore dock at Cleveland, Ohio.—Tom Hollyman, Hagley Museum and Library.

4
Yards, Docks, and Terminals

The transportation of freight has always been the primary business of American railroads. The railroad brought an unparalleled efficiency to the overland movement of freight, and an evolving railroad system transported the agricultural products, fuel, raw materials, and finished manufactured goods that supported and sustained the American industrial revolution. As they sought more efficient ways to handle this ever-increasing freight traffic, the railroads built some notable facilities to do the job.

From the very beginning, the classification yard has been an almost universal freight handling facility. Inherent in railroad technology is an ability to assemble individual "retail" carloads of freight with diverse origins and destinations into "wholesale" trainload quantities for efficient transportation. The key to this capability is the classification yard, where these diverse carload shipments are sorted and assembled in proper order into trains for movement over a common route, or trains are disassembled and their cars reorganized for delivery or the next stage of their journey.

"A classification yard," wrote John A. Droege in his *Freight Terminals and Trains*, "is essentially a machine for separating trains or drafts of cars, according to prearranged plans for distributing the cars in groups according to destinations, routes, commodities or traffic requirements, so as to accomplish their movement to tracks for these purposes."

In addition to the classification yard itself, the facilities of a freight yard typically include arrival or receiving yards to handle incoming trains and advance or departure yards in which departing trains are assembled from the classification tracks. At smaller freight yards the receiving, classification and departure yards usually handle traffic moving in both directions, while at larger installations separate yards are often provided for each direction of traffic.

Originally, classification yards were simple "flat" facilities, typically made up of parallel body tracks connected successively by one or more ladder tracks. Classification was accomplished by "push-and-pull switching" or "tail switching," in which cars were switched to the proper classification track by a locomotive. While many smaller classification yards still operate in this manner even today, it was never a very efficient procedure, and railroads were looking for a better way to classify cars well before the end of the nineteenth century.

One of the earliest alternatives to push-and-pull switching is a now largely forgotten switching method called "poling," which eliminated a large part of the back and forth movement that was required with conventional flat switching. It involved the use of a poling engine operated on a track parallel to that occupied by the cars to be separated or grouped. A diagonal wooden pole attached to the breast beam of the locomotive was placed in contact with a poling pocket on the corner of the car, or the last car in a cut, to be moved; and the

The development of the pneumatic retarder during the 1920s was a key step in the evolution of the modern classification yard. A car passes through a retarder at the Milwaukee Road's Pigs Eye Yard at St. Paul, Minnesota.—William D. Middleton.

The advanced classification yard completed by the Santa Fe at Barstow, California, in 1975 employs this massive master retarder at the top of the hump.—William D. Middleton.

engine then was used to start the cars. Once in motion, cars were switched into the proper classification track. Sometimes a poling car was used, with two poles on each side, one facing in each direction, that permitted an engine to pole cars in either direction along parallel tracks on each side of the poling track. Savings of as much as 40 percent in switching mileage and 20 percent in overall costs were reported for one early poling yard. But poling was hard on equipment and dangerous to switchmen if a pole jumped out of its socket under pressure, and the method eventually fell out of favor.

A much more satisfactory method for improving the efficiency of classification yards was the use of gravity switching, which employed an incline and the assistance of gravity to move cars in a classification yard. This was sometimes done by building a yard on a site at which a natural incline could be used, but more often it was accomplished by grading an artificial summit or hump above the level of the yard to provide the desired incline. The first of these hump yards in North America was the Pennsylvania Railroad's Honey Pot Yard near Nanticoke, Pennsylvania, built in 1891.

Hump yards use a locomotive to push a train over the hump at a low speed. At the crest of the hump, cars are uncoupled singly or in cuts as required for classification and then roll down the hump grade to be switched into the appropriate classification track. At first, switching was done manually by switchmen, but by the turn of the century this was often controlled by a single operator through the use of electro-pneumatic machines. These were usually installed in a tower at the hump where the operator could see both car markings and the position of the switches leading into the classification yard.

While the hump yards greatly increased classification productivity, they were still labor intensive facilities. In these early yards large numbers of "car riders" or "car droppers" were required to ride each cut or car down into the classification yard to set the brakes and control the speed of the cars. This was needed to stop a car at the desired location and to prevent damage by excessive coupling speeds with cars already in a classification track. In most yards, the car riders then walked back to the hump for the next cut. Sometimes a locomotive and car or a gasoline speeder were used to return car riders to the hump more rapidly. A few yards even had electric trolley lines to provide a fast return trip for hump riders.

One of the greatest single advances in hump yard technology came during the 1920s, when the development of pneumatic retarders provided a way to control car speeds without putting a rider on each car to set the brakes manually. These consisted of a number of cast iron shoes placed along both sides of each rail and operated by compressed air from a centrally controlled plant to grip both sides of the car wheels to reduce speed. A car usually passed through two or three sets of retarders as it progressed down the hump into the classification yard.

The first North American use of retarders was in 1924 for the Indiana Harbor Belt's hump yard at Gibson, Indiana, where they were installed on the yard's north hump. The installation eliminated the need for 60 car riders, and *Railway Review* reported that retarders had increased the yard's daily classification productivity from 25 to 43 cars per man.

The highly efficient retarders pioneered by the Gibson yard were widely adopted for the operation of larger classification yards all over North America and have remained a basic component of standard hump yard technology ever since. Even more recently the development of highly automated hump classification yards has

The Pennsylvania Railroad's grain elevator at Baltimore, Maryland, was typical of the massive facilities built by railroads for this bulk traffic.—Hagley Museum and Library.

enabled railroads to further reduce their yard costs and move freight cars through yards more rapidly, while the more accurate control of retarders provided by automatic operation has dramatically reduced the loss and damage from rough coupling that came with errors in judgment by retarder operators.

This shift to automated yards began shortly after the end of World War II with the application of electronics, telecommunications, and computers to management of the classification process. The first yards to apply the new technology were the Santa Fe's Argentine Yard and the Rock Island's Armourdale and Silvis yards, all completed in 1949. By 1960 more than 50 of these automated yards had been developed in the United States and Canada, either as entirely new facilities or through modernization of existing yards, and many more have been added since then.

In these modern yards, hump engine speeds are automatically controlled by computer. Based upon train consist data provided from management information systems, computers assign classification codes to each car being humped, and switches are automatically set to route each car or cut into the proper track. Computer-controlled retarder installations automatically set retarders to attain desired classification track coupling speeds. The data used to do this include weight and other car data, wheel sensor "rollability" measurements that are automatically generated as each car approaches retarders, radar speed measurements, and distance-to-coupling measurements. Once classification is complete, computers automatically generate switch lists and consists for outbound trains.

A good example of these modern automated yards is Union Pacific's Bailey Yard at North Platte, Nebraska. Now recognized as the world's largest rail yard, Bailey spreads out over 2850 acres. On an average day the yard handles 120 trains, classifies 3000 cars, and moves a total of 10,000 cars. Cars move over its two humps at a rate of four a minute into a 64-track eastbound classification yard and a 50-track westbound yard, all of it controlled from a computer-based yard command center.

The unloaders developed to empty box cars of grain were mechanical wonders. This shows the grain unloader at work at the Great Northern Elevator at Superior, Wisconsin, in 1957.—William D. Middleton.

But even as classification yards have gained in efficiency, reducing the unproductive time spent by cars waiting in yards, railroads have been devising ways to avoid using yards altogether. Container and trailer trains typically run through between intermodal terminals, while such measures as run-through trains, pre-blocking, or block swapping have helped to limit or eliminate yard time for carload traffic. Increasingly, what classification work remains is being concentrated in a lesser number of larger and more productive automated yards.

The efficient transportation of bulk commodities has always been the railroad's strong point, and the requirements of trans-shipping prodigious quantities of grain, ores or coal have generated a need for some remarkable cargo-handling facilities.

The western prairies of the United State and Canada produce an enormous flood of upwards of 400 million tons of wheat, corn and other grains every year. From track-side elevators on the prairies this great flood of grain flows in box cars or specialized grain hoppers to the milling centers or to the ports of the Great Lakes, the Gulf or the two oceans for export. Here, it is loaded into great elevator storehouses where it may be cleaned, dried, mixed, and stored until it goes to the mill or is loaded into ships for export.

These terminal elevators are often enormous structures. A good example of these massive elevators is the now-closed Great Northern Elevator on Lake Superior at Superior, Wisconsin. With a total storage capacity of more than 12 million bushels, this was one of the largest elevators in the United States. During a typical year in the mid-1950s, the elevator unloaded 17,047 cars carrying 30 million bushels of grain and loaded 135 lake steamers. On its biggest day that year the elevator unloaded 267 cars with 449,000 bushels of grain.

These electrically operated coal car dumpers used by the New York Central and the B&O at a joint coal and ore facility at Toledo, Ohio, were capable of dumping a car every minute. The ship is the *Howard M. Hanna, Jr.* of Wilmington, Delaware.—Baltimore & Ohio, David P. Morgan Memorial Library.

This was a typical upper Great Lakes ore loading dock. The Interlake Steamship Company's ore boat *C. H. McCullough, Jr.* is loading from a timber Duluth, Missabe & Northern dock at Duluth, Minnesota, in 1909. —Louis P. Gallagher, Northeast Minnesota Historical Center (Collection S2386).

Most cars arriving at the elevator were unloaded on an automatic car dumper which could unload a car in about 7 minutes and handled as many as 108 cars in a 13-hour workday. This car dumper was mounted on a large cable-driven trunnion which could tilt a car 35 degrees in each direction, while the table on which the car was placed could tilt 15 degrees to one side to help empty the car. From a hopper under the dumper, the grain was conveyed to an overhead "garner" and thence to scales, or for drying, cleaning or mixing, before finally moving to storage bins for eventual loading into lake steamers or rail cars.

The development of the steel industry around Pittsburgh in the last half of the nineteenth century generated a demand for enormous quantities of iron ore. This ore traveled from the iron mines of Upper Michigan and Minnesota largely by Great Lakes ore steamers, but it began and ended its trip by rail, necessitating some extraordinary structures to transfer the heavy ores from rail cars to ships and then from ships back to rail cars.

A typical loading dock design evolved early in the ore ports of the upper Great Lakes. Initially built of wood, and later of steel and concrete, these were typically high, gravity-operated structures with long approach trestles or viaducts, onto which trains of ore cars were pushed. These were then unloaded through bottom dump hoppers into a series of hundreds of "pockets" on the dock, each capable of storing the ore from several cars. When a ship was ready to load, these were then emptied into the vessel by means of long gravity chutes.

Unloading of ore boats in the lower Great Lakes was a much more difficult process. Originally it was accomplished largely by manual labor, but this was a costly and time-consuming process; and a variety of mechanical ore unloading machines were soon developed. Among the earliest types were machines that traveled up and down a dock on their own tracks, employing clamshell buckets to remove the ore through hatches on the ore boat. The clamshell bucket could then be moved back from the dock face on a traveling carriage to be unloaded into rail cars on tracks beneath the machine or storage locations.

An example of these was the B&O's ore unloading dock at Lorain, Ohio. This was equipped with three enormous electrically driven Brownhoist machines with grab buckets that could scoop up 7 to 10 tons of ore on each trip into an ore boat hold and then dump it into rail cars or storage hoppers. The installation was capable of unloading a thousand tons of ore an hour.

Still more advanced was the Hulett Automatic Unloading Machine developed in 1899 by George H. Hulett, an engineer from Conneaut, Ohio, who had specialized in designing bulk material handling equipment. Hulett's wonderfully ingenious machine consisted of two parallel girders at right angles to the length of a dock, supported by columns fitted with trucks so that the machine could be moved on rails to any point along the dock face. The two girders supported a carriage or "trolley," which, in turn, carried a walking beam. The inner end of the walking beam was equipped with a hoisting mechanism which could rock the beam, while the outer end supported a leg equipped with a large bucket at its lower end. This bucket leg was suspended in a vertical position and was mounted on rotating trunnions in the walking beam so that the bucket could rotate 360 degrees around its vertical axis and reach out in any direction beneath an ore boat hatch, while it could be moved laterally across the width of the vessel.

An operator located in a small cab inside this vertical leg, just above the bucket, rode up and down with each load and was able to see and control the operation of the bucket at all times. By means of a hoisting mechanism, the bucket was lowered through a ship's

The worldwide shift to containerized transportation required the development of new types of equipment for the intermodal terminals at which the containers made the transition from one transport mode to another. This is a rubber-tired Transtainer gantry crane used to load and unload container rail cars at the newly completed Wolfe's Cove terminal for Canadian Pacific's transatlantic container traffic at Quebec City in March 1971.—CP Ships.

hatch to the ore in the hold, where the bucket was closed and filled. Then the leg was raised and carried back over the dock by the trolley and the bucket discharged into a hopper. While the bucket was returning to the ship for another load, the ore in the hopper was dumped into an auxiliary "bucket car" or "larry" which weighed the ore and then dumped it into a storage area or rail cars on tracks below the Hulett machine.

A typical Hulett unloader had a capacity of about 10 tons of ore and could complete about one unloading cycle a minute. Speeds of as much as 60 buckets in 40 minutes were sometimes attained, and production rates as great as 783 tons an hour were reported for a single Hulett machine. The machine's great flexibility enabled it easily to reach both sides of an ore boat hold and into the area between hatches. On some ore boats the Huletts were able to remove as much as 97 percent of the ore before hand labor was required to fully empty a hold.

Hoisting, lowering, rotating, and trolleying on early Huletts were accomplished by hydraulic cylinders, with the power supplied from a steam-powered hydraulic pump. A separate auxiliary steam engine was used to move the machine along the dock and to operate the bucket car. Later versions of the Hulett unloader were electrically powered. Altogether, some 77 Huletts were built over a 50 year period for ore unloading docks on the lower Great Lakes.

Over the last few decades a shift to much larger ore carriers, together with the adoption of advanced bulk material handling systems, has transformed the way ore is loaded and unloaded at the Great Lakes iron ore ports. Fast conveyor belt ship loader systems have proved a quicker and more efficient way to load the ore boats than the traditional gravity pocket system. Ore carriers are now almost always equipped with their own self-unloading systems, ending the need for such big dinosaurs as the Hulett unloaders or similar systems. The old existing docks are being adapted to these new systems, and new docks are likely to be entirely different from the older facilities they replace.

Coal has long been the most important single commodity transported by American railroads. Today U.S. railroads transport more than 700 million tons yearly, representing some 44 percent of their total freight tonnage. A sizable part of this traffic flows from the great coal fields of Pennsylvania, Maryland, Virginia, West Virginia, and Kentucky to the tidewater ports of the eastern seaboard for export or coastal shipment. At the ports where this enormous traffic is transferred from rail cars to ships, railroads have developed some enormous and productive coal piers that make this transfer rapidly and economically.

The earliest coal piers consisted of simple elevated structures. The coal cars were pushed out along the top of the pier, and their loads were discharged either through chutes to the waiting ships or into pockets for later loading. Typically there was a track for loaded cars on each side of the pier, with a center track for the removal of empty cars.

Later, a variety of mechanized coal pier designs was developed. Some of these employed giant car dumpers which could pick up loaded cars, turn them over and dump their contents into a pan or chute which then conveyed the coal into a telescoping chute for placement in the hold of the ship being loaded.

A coal pier of this type built in the early 1920s at the Reading's Port Richmond Terminal on the Delaware River at Philadelphia, for example, could handle cars

Another view of the then-new Wolfe's Cove intermodal terminal shows one of the straddle carriers used to move containers in the terminal and the 40-containers-an-hour Portainer crane used to load ships.—CP Ships.

weighing as much as 165 tons, car and contents, at a rate of about 40 cars an hour. Loaded cars were pulled up to the dumper on an inclined trestle by a "barney" car with a swinging arm which rose up to engage the coupler of a car while pushing it up to the dumper cradle. Once emptied, cars were returned by gravity to a switchback at the end of the pier and then to a storage yard for empty cars.

The B&O built a coal pier at Curtis Bay, Maryland, near Baltimore, that operated in a much different manner. Capable of loading 12 million tons of coal a year aboard ships, the Curtis Bay pier employed two car dumpers at the land end of the pier, each capable of dumping 40 cars an hour. Loaded coal cars were dumped into a huge "pan," which then deposited the coal into hoppers, from which it was transported by conveyors to storage bins at the inner end of the pier. Coal was then fed from these bins by high-speed conveyor belts to big loading towers on the pier that could each load 2000 tons an hour into a ship.

Still another arrangement was devised by the Virginian Railway for its first coal pier at Sewalls Point, in Norfolk, Virginia. A rotating car dumper emptied loaded cars into big electrically powered conveyor cars, which were then raised to the top of the pier either by an elevator or by hauling up an incline. The conveyor cars then operated under their own power up and down the top of the pier to dump their coal into pockets on either side of the pier from which it was discharged through chutes to the holds of coal ships.

In the second half of the twentieth century few shifts in technology have been more important to the railroad industry than the development of intermodal freight transportation. While the idea goes back much further, intermodal first began to take off in the 1950s with the advent of trailer-on-flat-car services, and it has been the fastest growing segment of rail freight traffic ever since. By 1996, U.S. railroads were moving well over 8 million trailers and containers every year, and an entirely new type of terminal had been developed to handle the intermodal transfer between road and rail carriers.

If intermodal freight transportation has changed the railroad industry, it has transformed ocean shipping. Manufactured goods and commodities of every description that were once carried in the holds of cargo ships now move on board dedicated container ships in standardized containers. Railroads have been playing a growing role in international shipping ever since the early 1980s, when they pioneered the development of double-stack container cars and the concept of dedicated "land bridge" intermodal trains linked with ocean carriers. Speed is of the essence in transporting this high value containerized freight. To provide it, steamship lines have developed enormous, fast container ships that can transport as many as 6000 containers at speeds up to 25 knots, while the railroads operate their intermodal "land bridge" trains at near passenger train speeds.

A key element in the success of this fast sea-land service has been the development of an entirely new kind of seaport intermodal terminal that can make a swift hand-off between ship and rail car. Highly mechanized, these terminals employ giant high speed cranes, sophisticated container-handling equipment, and computerized control systems to develop an extraordinary productivity that can send containerized cargo on its way rapidly and efficiently.

The Bessemer & Lake Erie's Ore Unloading Dock No. 4 (1899) at Conneaut, Ohio

Late in the nineteenth century Andrew Carnegie assembled a vast, integrated steel-making enterprise organized around his Carnegie Steel Company. Carnegie gained control of coal mines, iron mines on Minnesota's Mesabi Range, and a transportation chain of the railroads and Great Lakes ore boats needed to ship raw materials to his Pittsburgh area steel mills. The Bessemer & Lake Erie formed a key link in this enterprise between Lake Erie and Pittsburgh.

A B&LE predecessor had reached the Lake Erie port of Conneaut, Ohio, in 1892; and by the end of the year the railroad had completed the improvements needed to handle ore boats. Initially, only a single timber pier was available but the railroad added two more during 1897-98 to handle both the growing ore traffic and outbound coal shipments.

Ore traffic through Conneaut grew still more as Carnegie-financed improvements were completed on the B&LE's route between the port and Pittsburgh. Construction began in 1898 to widen the harbor and build another ore dock on the east side of the harbor opposite Dock No. 1. Completed the following year, Dock No. 4

An early view of Dock No. 4 shows the battery of steam-powered Hulett ore unloaders that were installed beginning in 1899.—Pittsburgh & Conneaut Dock Company.

This view of the construction of the first steam-powered Hulett ore unloader at Conneaut's Dock No. 4 around 1899 provides a good view of the unloader's novel walking beam arrangement and the vertical bucket leg that reached down into the hold of an ore boat.—Pittsburgh & Conneaut Dock Company.

was a 1775-foot-long structure of timber construction that would become Conneaut's principal ore-handling facility.

Ore unloading equipment for the new dock included a battery of Brownhoist "fast plant" clamshell unloaders, but a group of George Hulett's automatic unloading machines soon became the heart of Dock No. 4 operations. This first Hulett unit, completed in 1899, was a 15-ton capacity steam-hydraulic machine; and it proved so successful that three more were installed over the next two years. In addition to the unloading machines, the pier was equipped with a long, track-mounted ore bridge fitted with a traveling clamshell bucket that was used to move ore to storage areas behind the pier and to later reload it into cars. The timber pier itself was rebuilt with concrete construction around 1907.

Still later, the railroad began to install electrically powered, 17-ton capacity Hulett unloading machines at the dock. The first went to work in 1910, and two more were added in 1916. The earlier unloading machines on Dock No. 1 were dismantled the following year, and Dock No. 4 then handled all ore shipments through Conneaut. Two more electric Hulett machines were installed in 1924, and the steam-operated Huletts were dismantled a year later. Each of the five big electric machines was capable of lifting 17 tons of ore in a single bite, and they could unload 3750 tons of ore in an hour. Typically the Huletts unloaded a vessel holding anywhere from 10,000 to 18,000 tons of ore in 3 $\frac{1}{2}$ hours. Ore was discharged from the unloader into a hopper on the Hulett frame and then transferred to a "larry" conveyor car for weighing and dumping into either hopper cars or a pit from which clamshell buckets on the big ore bridge moved it to a storage area.

For close to 60 years the big electric Huletts remained the backbone of ore-handling operations at Dock No. 4, and they established some impressive records as both the number and size of Great Lakes ore boats grew, together with steadily increasing ore tonnages. What may have been an all-time record for the Hulett unloaders was set on June 6, 1943, when the five machines lifted 14,275 tons of ore from the steamer *D. G. Kerr* in just 2 hours 45 minutes, an average of 5192 tons an hour. Dock No. 4's biggest year ever came in 1953, when a record 13 million tons of ore moved through the port.

Changes began to come to the B&LE's dock operations in the years after World War II. To provide a reserve supply, as well as to stockpile various grades of ore, the storage area behind the dock was enlarged to cover a 13-acre site with a capacity of some 1.4 million tons of ore. Further changes appeared in the 1970s with a shift to larger, self-unloading ore boats on the Great Lakes. As the number of self-unloading vessels grew, use of the Huletts became less frequent. The big machines finally unloaded their last vessel in 1985 and were dismantled in 1994.

Late in 1972 the B&LE completed the installation of an entirely new, highly mechanized ore handling and storage system for the new self-unloading vessels that can handle up to 10,000 tons of ore an hour. Unloaders on the vessels dump the ore into hoppers which feed a big 72-inch conveyor belt installed along the dock face. This transports the ore to a transfer station, from which it is carried by a second conveyor to an open storage area on a hillside behind the dock area. Here, on a 58-acre site, the railroad can store as much as 3 million tons of ore until it's needed by the steel mills.

When the time comes to load hopper cars for shipment to the mills, the ore is reclaimed from the storage area by two huge track-mounted reclaim wheels and deposited back on the conveyor belt for movement to a car-loading facility. Here, the ore is weighed and dumped into hopper cars which are moved into position on tracks below the structure by a locomotive remotely controlled by the loading plant operator.

Now nearing its centennial, the B&LE's Dock No. 4 has successfully adapted to the industry's changing needs to remain an important link in the vast enterprise of

This close-up view of the vertical bucket leg of one of the Hulett unloaders at Dock No. 4 shows the operator's position just above the bucket.—Pittsburgh & Conneaut Dock Company.

When this photograph of the ore boat *Percival Roberts, Jr.* was taken at Dock No. 4 around 1924, five electric Hulett unloaders, to the right, and three older steam-powered Huletts worked to unload the vessel.—Pittsburgh & Conneaut Dock Company.

Yards, Docks, and Terminals   **149**

By the time this view of Dock No. 4 was taken in the early 1980s self-unloading vessels and conveyor systems had taken over much of the work of unloading ore carriers. The last four Huletts and the big ore bridge seen here were dismantled by the mid-1990s.—Pittsburgh & Conneaut Dock Company.

Iron ore or taconite pellets shipped through Conneaut's Dock No. 4 are likely to end up in this big storage area behind Dock No. 4 until it's time to ship them to Pittsburgh-area steel mills. This track-mounted ore reclaimer scoops up the stored ore and sends it by conveyer belt to the ore car loading facility.—William D. Middleton.

steel-making. Ore traffic through Conneaut is down from the peak years of Great Lakes ore shipping, but anywhere from 2 to 3 million tons of ore still moves across the venerable dock every year.

GETTING THERE

Conneaut is just 2 miles west of the Pennsylvania state line on U.S. 20 and can be reached from Exit 241 on I-90. The harbor and dock facilities are located a short distance northeast from the downtown center.

The Pennsylvania's Enola Yard (1906) at Harrisburg, Pennsylvania

An important part of the Pennsylvania Railroad's massive improvement program begun in 1900 was an enormous new freight classification yard at Harrisburg, Pennsylvania. This was at a key junction of the railroad's east-west main line with the affiliated Northern Central and the western end of a new low-grade freight line that bypassed the Philadelphia area.

The new Enola ("alone" spelled backwards) yard was built on the west bank of the Susquehanna, just south of Marysville and the Rockville bridge. At its north end Enola was linked with the main line and the Northern Central line to the north, and at its south end to the Northern Central's Baltimore line, and to the new low-grade line, which came across the Susquehanna to the Northern Central at Shock's Mill.

The enormity of the Pennsylvania's Enola Yard is evident in this northward-facing aerial view. The eastbound classification yard is in the foreground. Beyond are the car shop, engine terminal, and westbound receiving and classification yards, while the Susquehanna River is to the right.—Tom Hollyman Aerial Photographs, MG-286 Penn Central Railroad Collection, Pennsylvania State Archives.

Yards, Docks, and Terminals

Above: a southward aerial view of the yard shows a train being classified on the westbound hump, with the engine terminal beyond. The Susquehanna is to the left.—Tom Hollyman Aerial Photographs, MG-286 Penn Central Railroad Collection, Pennsylvania State Archives. *Right (both photos):* Taken November 19, 1952. Above, eastbound hump classification yard at night. The retarders in the foreground were controlled by an operator in the tower on the right. Below, 2-10-0 No. 4432 pushing the last car of a cut over the eastbound hump. —Philip R. Hastings, David P. Morgan Memorial Library.

—Drawing by Christian Goepel.

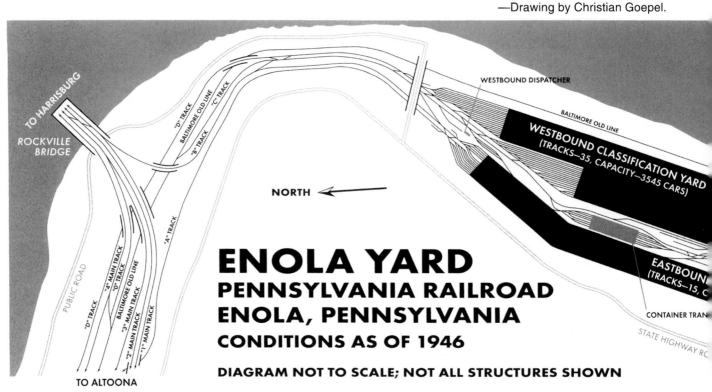

ENOLA YARD
PENNSYLVANIA RAILROAD
ENOLA, PENNSYLVANIA
CONDITIONS AS OF 1946

DIAGRAM NOT TO SCALE; NOT ALL STRUCTURES SHOWN

Over a three-year period from 1903 to 1906, construction crews graded the huge site, laid miles of track, and erected the yard's extensive support facilities. The yard spread out for miles alongside the river, with separate receiving yards, humps and classification yards for eastbound and westbound trains. It was said to be the largest yard anywhere in the world.

Enola became the principal yard for the eastern end of the PRR system, and it remained in a more or less continuous state of growth and change as the railroad added or improved facilities to cope with traffic growth. A major improvement came in 1938, when electropneumatic retarders began operating on the yard's eastbound hump, cutting classification time by 20 percent and displacing the car riders that had ridden cars down the hump since 1906.

At the beginning of the 1940s Enola was the largest and most important classification yard on the entire Pennsylvania system. One PRR official likened it to "the throat through which very nearly all of the Pennsylvania's through traffic must pass."

Within Enola's yard limits there were 145 miles of track and 476 switches. The entire yard had a standing capacity of close to 12,000 cars. For fast freights and coal trains from the west and north there was a 14-track eastbound receiving yard from which traffic either went over the hump into the 33-track classification yard or went straight into a separate six-track solid train yard. Departing trains for the east and the south were made up in a five-track eastbound advance yard. Westbound traffic came up the west bank of the river to enter a 16-track receiving yard, from which it moved over the hump or directly into a 35-track westbound classification yard. Westbound trains departed directly from the classification yard.

At the center of the yard was a transfer facility for the Pennsylvania's pioneering container system for less-than-carload-lot traffic. An icing plant serviced refrigerator cars. Separate car repair yards supported both the eastbound and westbound classification yards, while a car overhaul shop and paint shop turned out dozens of rebuilt and repainted cars every day. A 42-stall roundhouse, two turntables, engine storage tracks, and coaling and watering facilities at the center of the yard supported more than a hundred steam and electric locomotives dispatched from Enola every day.

As the buildup to World War II began, Enola was put to its severest test ever as traffic set one record after another. Traffic grew rapidly from a 1939 average of 11,207 cars handled daily. In October 1941 the yard recorded a total of 437,127 cars—a daily average of over 14,100 cars—passing through the yard. Total traffic for the year was well over 4.8 million cars. This reached more than 6 million in 1943, the yard's biggest year ever. Enola set a one-day record on June 19, 1943, when 20,660 cars went through the yard. In 1943 Enola's engine terminal was dispatching a daily average of 160 steam and electric locomotives.

Somehow Enola weathered the storm, but it was a close call. "During the past winter," wrote a yard official

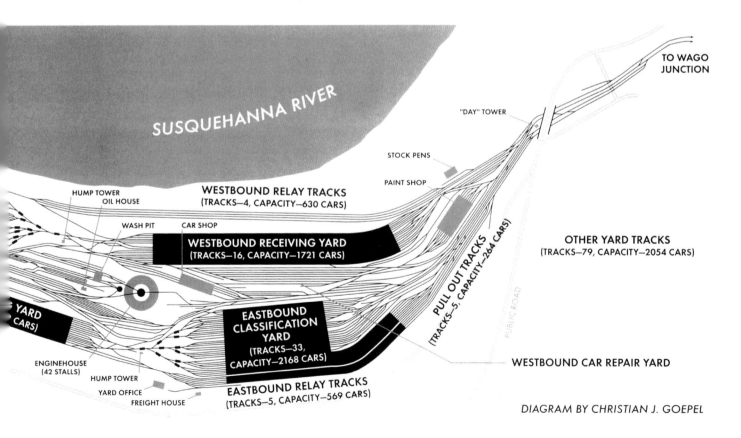

DIAGRAM BY CHRISTIAN J. GOEPEL

in 1943, "there were several occasions when a near-blockade of traffic through Enola Yard existed, caused in a large measure by inability to turn out engines fast enough to take trains from the yard as they were made up."

Still more improvements helped Enola to remain fluid during the war years. Chief among these was a modern retarder installation completed in late 1944 that finally replaced car riders on the westbound hump.

Enola remained an important yard for the Pennsylvania well after the war's end. Trains were growing longer with diesel power, and Enola's yard tracks were extended to accommodate trains of as many as 144 cars. The yard's total capacity had grown to 12,888 cars on 192 tracks. As late as 1953 the yard was handling an average of 11,000 to 12,000 cars daily, and the yard's three-shift daily work force totaled 4000. Enola's engine terminal was the home base for 248 diesels, 90 electric road units and 90 steam locomotives.

But as freight traffic began to decline in the great industrial heartland served by the Pennsylvania, Enola's fortunes faded with it. In 1957, when the railroad decided to establish its first automated yard, the investment went to Conway yard, west of Pittsburgh. This made Conway the Pennsylvania's principal yard, taking from Enola such functions as westbound classification. Enola never did get a modernization investment; and Conway continued to grow, displacing Enola from its "largest freight yard" claim by the end of the decade.

Further decline came with the shifting traffic patterns that followed the 1968 Penn Central merger and the formation of Conrail in 1976. Before the end of the 1980s, Conrail had shifted much of its freight traffic off the former Pennsylvania lines to the east. Most traffic to and from the east now moved over former Reading and Lehigh Valley lines, crossing the Susquehanna on the Rockville bridge and bypassing Enola; and the eastbound hump yard was shut down.

The end finally came for Enola in September 1993, when Conrail reorganized its freight operations and ended classification and most other functions at the yard. Little more than a diesel terminal and some car storage remained. After 87 years of operation, the yard fell silent on October 3, 1993, with closure of the westbound hump. Enola is still there alongside the Susquehanna, but it's no longer the largest and busiest classification yard in all of North American railroading.

GETTING THERE

Enola lies in West Fairview, on the west bank of the Susquehanna about 4 miles north of downtown Harrisburg and just south of I-81. Enola Road (U.S. 11 and 15) parallels the yard.

This construction view of the Duluth, Missabe & Northern's Dock No. 6 shows the massive steel columns and frames that support the ore pockets on each side of the dock.—Northeast Minnesota Historical Center (Collection S2386).

In this general view of the DM&N's Duluth ore docks dating to October 1919, the newly completed Dock No. 6 at the left and the older steel Dock No. 5 at the right flank the railroad's earlier timber facilities, Docks No. 3 and 4.—Northeast Minnesota Historical Center (King Collection S3742).

The Duluth, Missabe & Iron Range's Ore Dock No. 6 (1919) at Duluth, Minnesota

The greatest of all the iron ranges of the Lake Superior district is the Mesabi Range of northern Minnesota. Since the range opened in 1892, its rich iron mines have shipped close to 4 billion long tons of ore, and it's still coming at the rate of more than 40 million long tons a year.

The Duluth, Missabe & Northern, later merged into the DM&IR, completed its first ore dock on St. Louis Bay at Duluth in 1893. The big timber structure was typical of the docks that had become a standard for the upper Great Lakes. Ore cars were pushed up an approach to the top of the dock, where their contents was emptied into ore pockets, which were then emptied through chutes into the holds of ore boats.

Ore tonnages grew rapidly, and more big docks were added over the next two decades. World War I brought record levels of steel production, and in 1917 the railroad began construction of a new concrete and steel ore dock on St. Louis Bay that would rank as the largest ore loading dock in the world at the time of its completion.

This new Dock No. 6 was 2304 feet long, 76 feet 5 inches wide at the top, and stood 84 feet 5 inches high above the water level. Construction of the huge structure required over a million feet of timber piling, more than 174,000 feet of 12-inch steel sheet piling, nearly 60,000 cubic yards of concrete, and almost 30,000 tons of steel.

The foundation for the dock was constructed by driving steel sheet piling around the entire foundation area and pumping it full of some 182,000 cubic yards of sand, which was then leveled as the support for an enormous concrete foundation slab that was 2438 feet long, 69 feet wide, and 6 feet thick. Both the long approach viaduct and the dock itself were built with a structural steel frame. Once the foundation was complete, these were erected by a huge double traveler crane, which began working from the inner end of the approach, traveling atop the completed structure as the viaduct and then the dock were erected ahead of it.

The 3487-foot approach was made up largely of a double-track steel viaduct, while the dock superstruc-

Ore loading went on around the clock at the DM&IR's big Duluth docks. The Shenango Line's ore boat *Schoonmaker* loaded at Dock No. 6 on the evening of May 25, 1959.—William D. Middleton.

emptied into the pockets. Once a vessel was docked and ready to load, a long spout on each pocket was lowered to dump the ore into the vessel.

Dock No. 6 was ready to load its first vessel in 1919. By this time the ore moving through the Duluth docks had begun to drop off from the level of the peak wartime years. The last of the DM&N's old timber docks were removed by the end of the 1920s, and Docks No. 5 and 6 handled all ore shipments through Duluth thereafter. After the slump in traffic of the depression years, ore tonnages began to rise again as U.S. industrial activity picked up to supply the materiel needed to fight World War II. In 1942 the two docks loaded what would prove to be their greatest annual traffic ever, when just under 24 million long tons of ore were handled.

Ore tonnages remained high well through the 1950s, but since then change has come to the way iron ore is processed and shipped. The Missabe Road's massive Dock No. 6 has proved remarkably adaptable to the new era.

The first big changes to the dock came during 1964-65 as the range shifted to the production of ore from taconite. In order to keep taconite processing plants running throughout the year, the railroad added a conveyor system to the dock and a 2.5 million ton taconite storage area next to it. During the winter, when the lakes are closed to navigation, ore shipments are transferred from the dock to the storage area through specially modified ore pockets and the conveyor system. The process is reversed during the shipping season, when big bucket wheel reclaimers reload the ore pellets from the storage area and feed them back through the conveyor system to the dock for loading into ships.

Another round of changes came to the dock during 1981–83, when the railroad modified it again to handle the larger ore boats becoming common on the Great Lakes and to add a fast, new shiploader system. A berth

ture was constructed with a steel frame and concrete partition walls, pocket walls and sidewalk slabs. Ore pockets were located the full length of each side of the dock. There were 384 of them and each could hold the contents of eight standard 50-foot ore cars, giving the dock a total storage capacity of 153,600 long tons. Two rows of steel columns, each designed to carry 1.5 million pounds, supported the ore pocket structure. The columns were spaced 12 feet apart and arranged in pairs to form bents under each transverse wall. The top of these bents was in the shape of an inverted V, forming a floor for the ore pockets that sloped toward the outside of the dock.

Loaded cars were moved onto four tracks on the dock, two over the pockets on each side, and then were

From a ship's deck the Dock No. 6 shiploader shuttles look like this in action. Each shuttle can load an average of 1350 tons an hour. This was the Cleveland Steamship Line's *Griffith*.—Skillings Mining Review.

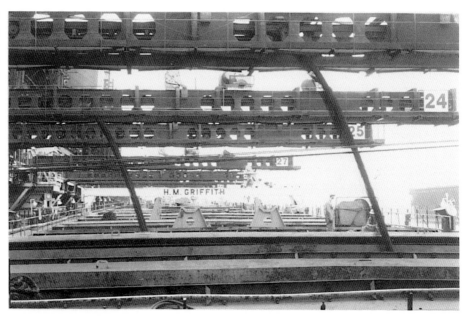

156 Landmarks on the Iron Road

In 1983 the DM&IR completed extensive modifications to Dock No. 6 for a high-speed shiploader system that could load a 60,000-ton-capacity vessel in only six or seven hours. The first vessel to load taconite at the modified facility was the massive *Columbia Star* on July 19, 1983.—Skillings Mining Review.

for these 1000-foot-long, 60,000 long ton capacity vessels was dredged on the west side of the dock, and a shiploader made up of 20 retractable shuttle conveyors was installed alongside it. These extend under the ore pockets on both sides of the dock, where they are supplied with ore from the pockets through a system of loading bins that discharge onto the conveyors. Each of the shuttle conveyors can feed ore into a ship's hold at a rate of 1350 long tons an hour, and the shiploader can load one of the massive new ore boats in only six to seven hours.

Dock No. 6 remains the largest ore loading dock of its kind ever built in North America; and with the changes of the intervening 80 years in ore boats and loading technologies, it is unlikely that anything like it will ever be seen again. And while it no longer equals the record tonnage figures of the past, the dock still loads a respectable average of some 7 million long tons of ore every year.

GETTING THERE

Dock No. 6 is on the north side of St. Louis Bay at West Duluth, which can be reached from downtown Duluth by Grand Avenue.

The Chicago & North Western's Proviso Yard (1929) at Chicago, Illinois

By any measure of size, the Chicago & North Western clearly laid claim to the title of "world's largest classification yard" when it opened the new Proviso Yard in 1929. It was one of the most advanced classification yards of its time as well.

Proviso was 14 miles west of downtown Chicago at the junction of C&NW's Chicago-Omaha main line with the low-grade Des Plaines Valley cutoff that linked it to the railroad's main lines to the north and northwest. This put it at the very heart of the North Western's network of lines that fanned out through the upper Midwest and linked the railroad with almost two dozen eastern and midwestern connections through the Chicago belt lines.

The North Western needed an improved yard to handle its heavy freight traffic to and from Chicago or its Chicago connections, and in 1923 the railroad began a long-term upgrade of an existing interchange yard at Proviso. These improvements would shift all of the North Western's Chicago terminal classification to Proviso, replacing the crowded Crawford Avenue yard 9 miles east of Proviso, which was hemmed in by urban development.

The Galena Division main line, which passed through the center of the old yard, was relocated to the south side of the yard; and a third track was added, extending from Elmhurst, at the west end of the yard, west to West Chicago. Much of the area for the new yard was swampland, and the railroad filled it with slag from Chicago steel mills, creating a site of some 1250 acres in an L-shaped area 5 1/2 miles long and a half-mile wide.

In a 1940s-era view of Proviso Yard, a cut of cars is pushed toward the hump from the North Yard receiving yard.—Chicago & North Western, David P. Morgan Memorial Library.

The general yardmaster at Proviso looked over his 260-track-mile domain from this observation and communications tower. A public address system carried his voice to all parts of the enormous yard. The photograph dates from the 1940s. —Chicago & North Western, David P. Morgan Memorial Library.

The new yard included the original inbound and outbound interchange yards, together with seven new yards. Eight of these were laid out generally parallel to the Galena Division main line. These included receiving, classification and departure yards for Chicago and connecting lines traffic, a classification and departure yard for Galena Division traffic, and another for the Wisconsin and Milwaukee divisions. Located on the north side of this complex were Proviso's twin humps, which sent cars down into a 59-track classification yard with a total capacity of 3300 cars. Car speeds into the classification yard were controlled by a total of 30 electric retarders. These retarders and the 58 switches that controlled car movements into the classification tracks were operated from three elevated control towers.

Along the north-south leg of the L-shaped yard, where traffic entered from the Des Plaines Valley cut-off, was the North Yard, a receiving yard for Galena and Wisconsin division traffic. Altogether, the yards that made up Proviso Yard had a capacity of 26,000 cars on 260 miles of track.

To the north of the hump yard was a 58-stall roundhouse and related locomotive servicing and repair facilities. Just inside the bend in the L-shaped yard was located an enormous 21-acre less-than-carload-lot freight transfer house that was said to be the largest building of its type in the world.

Still other facilities included three car repair yards, two scale tracks, office buildings, and an ice-making plant and icing facility for refrigerator cars. A communications system of telephone, teletype, two-way loudspeakers, and pneumatic tubes tied yard operations together and moved the paper. Powerful electric floodlights mounted on towers illuminated the yard for night-

In this aerial view of Proviso Yard from east to west, dating from 1959, the main classification yard is in the middle distance, to the right, with the hump in the distance. At one time in its history, Proviso had the capacity to store as many as 26,000 cars on 260 miles of track.—Chicago & North Western, Donald Duke Collection.

time operations. Overseeing it all was a general yardmaster located in a glass-walled control tower high atop the yard's ice plant.

Proclaimed the "largest individual freight terminal in the world," Proviso was officially opened on July 1, 1929. The economic depression of the 1930s that soon followed reduced freight traffic far below normal levels, and the enhanced capacity provided by the new yard would not be needed until the demands of World War II industrial activity hit the North Western and other American railroads. But as wartime traffic grew, business quickly picked up at Proviso. In just three years from 1940 to 1943; the North Western's freight tonnage increased by almost half. By October 1942 Proviso was classifying an average of 7600 cars every day, an average that went up to about 8000 per day later in the war.

Activity remained high at the yard after war's end. In a November 1949 profile in the railroad's *North Western Newsliner*, Proviso was said to be handling an average of 7000 cars daily, with an average of 128 trains

Yards, Docks, and Terminals

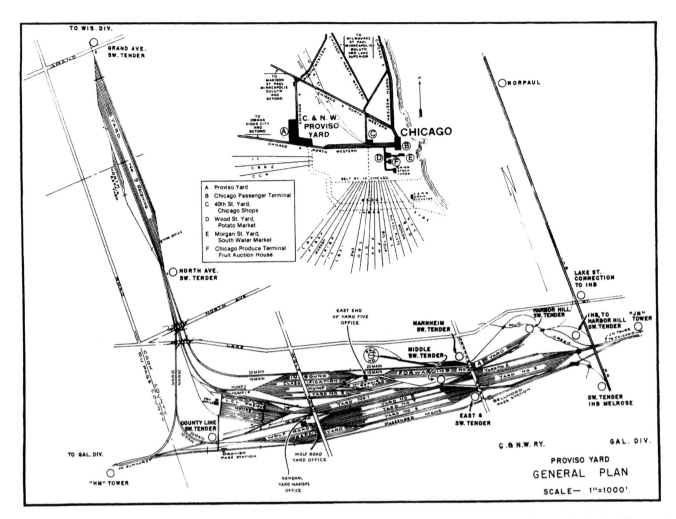

This general plan of Proviso Yard also shows diagramatically how the yard was linked to the North Western's principal lines and to its Chicago connections.—R. A. Janz Collection.

arriving or departing from the yard every day. A management force of a general yardmaster, a night general yardmaster, and 21 day, night and relief yardmasters supervised a total yard work force of 2200 employees.

By the late 1950s Proviso Yard was feeling the winds of change that a new management team had brought to a declining C&NW. For years a big sign on a nearby highway had proudly proclaimed that this was the world's largest freight yard. "Tear it down," ordered board chairman Ben Heineman. "We're not interested in the biggest—just the most efficient."

Efficiency meant moving cars through the yard faster, not storing them. Installation of a second lead for the Proviso hump increased classification capacity from 3000 to 4000 cars a day. This and other operating improvements enabled the railroad to close and remove three of Proviso's nine yards. In 1963 the railroad added a trailer-on-flat-car terminal for the Chicago area at Proviso that was capable of handling 700 trailers a day.

While its "largest freight yard" banner has long since moved elsewhere, Proviso is still an intensely busy classification yard and terminal. Since C&NW's 1995 merger into Union Pacific, Proviso has represented a major eastern connection and interchange point for UP, linking it with a dozen other Chicago railroads.

Today's Proviso Yard employs a work force of more than 1200. Its receiving yard includes 28 tracks with space for as many as 1400 cars, while its 42-track departure yard has an equal capacity. The 66 classification tracks fed by the hump have space for up to 1500 cars. The yard originates 28 to 30 west- and northbound manifest trains on an average day and handles anywhere from five to ten run-through trains. Proviso's modern "Global Two" intermodal terminal makes well over 200,000 container lifts a year, and it handles intermodal stack trains for three international container lines.

GETTING THERE

Proviso Yard is on the Chicago west side, between St. Charles Road and Lake Street (U.S. 20) and just north of Metra's West Line stations at Bellwood and Berkeley. North-south Wolf Road and Mannheim Road (U.S. 12, 20 and 45) pass over the yard on viaducts.

The Norfolk & Western's Coal Pier No. 6 (1962) at Norfolk, Virginia

The discovery of the rich Pocahontas coal fields in the mountains of southwestern Virginia in 1881 transformed the Norfolk & Western into a major coal railroad. The N&W began delivering coal to tidewater at Norfolk in 1883, and the Hampton Roads port has been a major coal terminal ever since. The railroad soon established a waterfront terminal at Lamberts Point, and its first coal pier was in operation by 1884. The coal traffic increased rapidly, and the Lamberts Point facilities grew, together with the business.

In 1960 the N&W was planning to build a new coal loading facility at Lamberts Point. By 1957 the railroad's export coal traffic had hit a record level of more than 19 million tons. Following its 1959 merger with the Virginian, the N&W was planning to shift all of its coal traffic to Lamberts Point. At the same time, huge new "supercolliers" of 100,000 tons capacity and more were beginning to operate in world coal trade that would require loading facilities of equally great capacity.

The new pier planned by N&W would be far larger and more productive than anything yet seen in Hampton Roads. Coal Pier No. 6 would be 1650 feet long and 82 feet wide, and it would have two huge traveling shiploaders that could load as much as 16,000 tons an hour.

The work began in 1961 with dredging of a pier slip 43 feet deep. About 1.5 million cubic yards of dredged sand and other material was pumped inshore as fill. This was used to create 29 acres of new land at the head of the pier and to bring the existing land area up to the level required for an enlarged yard for empty cars from the car dumping facility.

Once the fill was in place, the pier contractor began driving steel sheet piling to form a new bulkhead along the shore. Almost 2000 huge 24-inch square prestressed concrete piles, each designed to support an 80-ton load, were driven to support the new pier. About 175,000 linear feet of piling, in lengths of up to 95 feet, was brought by barge to the site, where a Vulcan pile driver drove it to depths of as much as 89 feet below mean low water.

The reinforced concrete pier is 1636 feet long overall and 82 feet wide. Two 6-foot-deep girders extend along each side of the pier. Cross beams spanning between these girders support a foot-thick deck slab. The two traveling shiploaders that move up and down the length of the pier are supported directly by the two outside girders.

These giant shiploaders are the principal elements of the pier's complex system of coal handling and storage facilities and the key to its ability to load rapidly the massive colliers that now call at Hampton Roads. Each of these 2800 ton towers is mounted on 96 wheels and travels along the pier on rails. They stand 192 feet above mean low water and are equipped with retractable booms with a reach of 120 feet and telescopic chutes that can load ships with a beam of up to 175 feet. The two shiploaders can either load two vessels at once or can team up to load a single ship.

Coal reaches each shiploader by a system of conveyor belts that can move 8000 tons an hour. This is fed from four high-speed rotary car dumpers that can empty

Made not long after the new pier began operating, this aerial view of Lamberts Point shows two ships being loaded from the big Pier No. 6 shiploaders. The older Pier No. 4 at the right was still active. Not yet in place were the two big surge silos at Pier No. 6 that were added in 1993 to substantially increase the pier's coal loading capacity.—Norfolk Southern.

The two enormous Pier No. 6 shiploaders dominate the Lamberts Point skyline. Each weighs 2800 tons, stands 192 feet above mean low water, and travels along its pier rails on 96 wheels. This pier level view of the outer shiploader dates to August 1997.—William D. Middleton.

up to 150 cars an hour into blending bins, which in turn feed twin 8-foot-wide conveyors that transport the coal to a transfer station and thence to the pier and the shiploaders.

Loaded cars are switched from the Lamberts Point coal storage yard into what's called the "barney yard" where they are staged to meet loading plans for a scheduled vessel. This yard is on a downward grade, and cars move from here to the dumper by gravity under the speed control of pneumatic retarders. On the way they first pass over electronic scales; and then, in winter weather, stop in one of two electrically heated thawing sheds before reaching the "barney pit," just before the dumper.

At the barney pit, cars are staged for movement to the dumper either singly or in pairs. A cable-powered "barney mule," operating on rails in a pit beneath the track level, rises up to engage the rear coupler and push the cars up an incline into the dumper. Once emptied, cars are ejected from the dumper by a mechanical device and move by gravity to a kickback incline, where they reverse direction to head off to the empty car yard.

Limited operation of the new pier began late in 1962, and N&W celebrated the completion of the world's largest and fastest coal loading facility on September 18, 1963. The splendid new facility then settled down to the business of loading large quantities of coal.

Several changes since it opened have given Pier No. 6 an even greater capacity. The berths alongside the pier and the access channel have been dredged to a depth of 50 feet to accommodate even larger colliers. In 1993 Norfolk Southern increased the pier's capacity by adding two big "surge" silos that can store 8000 tons of coal. These are linked to the conveyor system to permit car dumping to continue without interruption while ships move in or out of a berth or the shiploaders are being moved from one hold to another.

In the years since its completion, Pier No. 6 has set one record after another for loading speed and quantities. The pier has loaded a single vessel with as much as 153,000 tons of coal, and in one record 24-hour period in November 1990 the pier dumped 1772 cars containing 171,315 tons of coal. In 1990, the pier's biggest year yet, the big shiploaders handled an annual total of 39.5 million tons; and the pier's operators think they could get an annual capacity of 48 million tons out of the facility if they had to. With that kind of capacity NS closed its only other coal pier at Lamberts Point in 1989. Since then, Pier No. 6 alone has handled the railroad's Hampton Roads coal traffic, and it is still the biggest and fastest of its kind anywhere.

GETTING THERE

Pier No. 6 is at Norfolk Southern's Lamberts Point terminal, northwest from downtown Norfolk by way of Brambleton Avenue and Hampton Boulevard (State Route 337) and left on Redgate Avenue.

Both shiploaders were teamed up to load the collier *Cape Eagle* on August 6, 1997.—William D. Middleton.

A view of a Pier No. 6 shiploader from atop the outer shiploader shows the retractable boom that can reach out as much as 120 feet from each shiploader. Coal was being loaded through telescoping chutes into the holds of the *Cape Eagle* on August 6, 1997.—William D. Middleton.

The precise track geometry of a classification yard is evident in this night view of the classification yard "bowl" from the Spencer Yard tower on September 1, 1982. Cars are humped from the hump crest at the base of the tower through a master retarder, in the foreground, and then through one or more of eight group retarders further down the hump.—William D. Middleton.

The Southern's Spencer Yard (1979) at Linwood, North Carolina

The Southern Railway opened its first automated classification yard at Knoxville, Tennessee, in 1951. By 1973 there were six of these modern facilities at strategic points on the railroad, and they had helped it to post some notable gains in its overall operating efficiency and service reliability. By the mid-1970s the Southern was ready to add another.

The site for the new yard was determined by a system-wide study aimed at quantifying the potential earnings from a new yard and finding the best place to put it. Using data from the Southern's system-wide TIPS train information system, computer modeling studies simulated the traffic patterns that would result from different yard locations and calculated the effects on transit times, switching, and yard engine requirements at various terminals.

The studies pointed to a location somewhere near Spencer, North Carolina, as the one that would offer the greatest benefits. This would give the Southern its first automated yard anywhere on a 600-mile stretch of main line between Atlanta and Alexandria, Virginia. Nine of the Southern's ten most congested flat-switched yards

Yards, Docks, and Terminals

Photographed during humping operations on September 2, 1982, this view shows one car, in the left foreground, about to begin its trip down the hump. A second box car is just entering the master retarder, while a third, in the left distance, is well down one of the classification yard leads.
—William D. Middleton.

were located on its Eastern Lines, and five of them were relatively close to a North Carolina site.

Rebuilding the existing flat yard at Spencer would have required costly grade separations for four road crossings and would have created severe freight traffic disruption during construction. Instead, the railroad found an ideal location just 8 miles north of Spencer at Linwood, where the yard could be built without any interruption to traffic on a site free of road crossings.

The new yard was laid out just west of the Atlanta-Washington main line on a 376-acre site some 4 ½ miles long and 1700 feet wide at its widest point. About 5.6 million cubic yards of grading work were needed to prepare the site, and almost 8 miles of drainage pipe and nearly a thousand linear feet of triple box concrete culvert were needed to solve drainage problems. Almost a million tons of base material and 225,000 tons of granite ballast went into the yard, and more than 65 miles of track were laid with continuous welded rail.

The yard's layout and equipment are similar to its recent predecessors elsewhere on the Southern. An eight-track receiving yard at the south end is directly accessible from both north- and southbound main line tracks. From here, cars are humped from south to north into a

—Drawing by Christian Goepel.

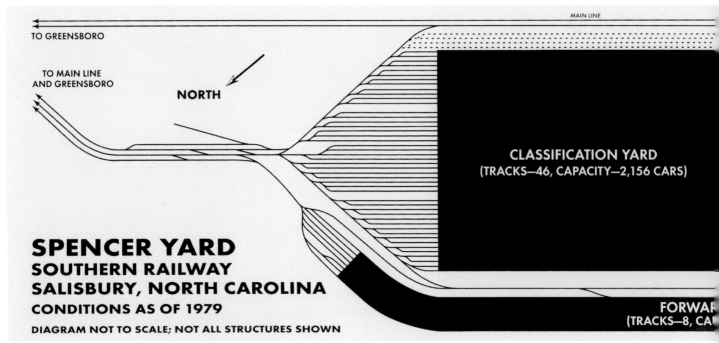

46-track classification yard that can hold a total of 2156 cars. Speed control is provided by an electric retarder installation, while weight-responsive "inert" retarders at the exit end of each classification track assure that a car does not roll out of the yard.

Three long pull-out tracks are located to the north of the classification yard, and both north- and southbound departing trains are made up in an eight-track forwarding yard laid out along the west side of the yard.

Other facilities include a five-track spot repair shop and an engine terminal located between the forwarding yard and the hump. Yard operations are managed out of a seven-story yard office and tower at the hump crest. A high intensity lighting system illuminates the yard for night operations.

But the real heart of this modern yard lies in its sophisticated microprocessor control system and its integration with the railroad's system-wide communication and terminal information systems. The automated yard control process begins long before a train pulls into Spencer's receiving yard. Consist and routing information are supplied by microwave from the railroad's TIPS main computer at Atlanta to the yard's management information system computers. A videotape made of each train as it enters the receiving yard is used to correct this data for any consist changes made en route.

As each car in a train approaches the hump, it is shifted from the yard's management information system computers to a process control system. This is a computer-based system that provides automatic switching and retarder control during classification. The installation's retarder control system employs car weighing, wheel sensors, wind velocity measurements, radar measurement of car speeds, and distance-to-coupling measurements down each classification track to provide a 4

In this September 1982 view of humping operations from a vantage point below the hump crest, the covered hopper in the foreground has just cleared the master retarder and has started down the center lead into Spencer Yard's classification bowl. At the crest of the hump another car has just rolled free from the train on the hump to begin its journey down into the classification yard. Operations in the highly automated yard are controlled from the tower to the left of the hump.—William D. Middleton.

mph or lower coupling speed in the classification yard. The computer also provides automatic hump engine speed control.

After a train has been humped, the process control computer generates an "as humped" switch list and pull-out work orders for the crews that move classified cuts to the forwarding yard to build an outbound train. As each

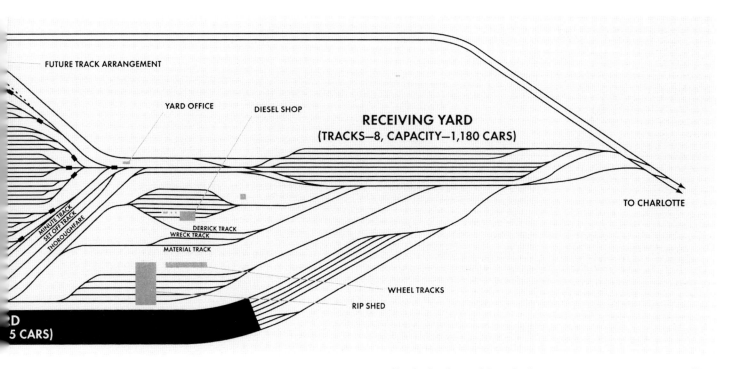

Yards, Docks, and Terminals

Spencer Yard's support facilities include an engine terminal and car repair facilities. This is a nighttime view of the locomotive servicing tracks in September 1982.—William D. Middleton.

of these is completed, the new location of each car is automatically reported to the yard's management information system computer. Once classification is complete with the departure of an outbound train, this in turn transmits the consist to the main TIPS computer at Atlanta, which can then update system car movement records and provide an advance consist to the next yard.

The opening of Spencer Yard in mid-1979 enabled the railroad to close its major flat-switched yard at nearby Spencer and to reduce the load at seven other, smaller flat yards in North and South Carolina. The new yard also helped the Southern to deliver improved service and to post some impressive cost savings. Average yard times were cut from about 30 hours in the old flat yard to only 14 hours, and freight transit times were reduced by as much as one to two days. Computerized retarder control cut car damage and lading loss and damage from excessive coupling speeds.

Merger with N&W in 1982 gave Spencer a changed role in freight classification as the new Norfolk Southern began to reroute traffic to get the best overall routing for the merged system. Today, Spencer Yard typically handles a daily average of anywhere from 20 to 25 inbound trains and builds a like number of outbound trains. Every day the yard typically humps around 1800 cars into the big classification yard, blocking traffic for such diverse terminals as Atlanta, Birmingham, Sheffield, Savannah, Asheville, Knoxville, Roanoke, Norfolk, and Hagerstown, or for its Conrail and CSX connections. Still the newest automated classification yard on Norfolk Southern, Spencer Yard remains an excellent example of modern, high-productivity automated classification yard technology.

GETTING THERE

Spencer Yard is located near Linwood, North Carolina, about 12 miles north of Spencer on I-85.

Deltaport Intermodal Container Terminal (1997) at Vancouver, British Columbia

A notable example of a modern seaport intermodal terminal is the Port of Vancouver's Deltaport terminal, which began handling trans-Pacific container traffic through the British Columbia port in June 1997.

Vancouver built the new terminal primarily to increase capacity over that of the port's two older container terminals on Burrard Inlet, opposite downtown Vancouver. It also allows the port and the two Canadian transcontinentals to take advantage of opportunities to compete for a greater share of trans-Pacific container traffic that now moves through U.S. ports. Vancouver, for example, is about a half-day's sailing time closer to the principal Asian container port at Hong Kong than is Los Angeles, while both CN and CP have established intermodal routings between Vancouver and Chicago that are more or less equal to those of BNSF and UP between Chicago and either Seattle-Tacoma or Los Angeles.

Deltaport's location at Roberts Bank on the Strait of Georgia puts it anywhere from one to two hours closer to the Strait of Juan de Fuca piloting stations than either of Vancouver's older terminals, or the ports of Seattle and Tacoma. A 40-mile BC Rail line affords a direct access to the CN and CP main lines that entirely bypasses yards and congestion in the Vancouver area.

The terminal was built on a 100-acre site on an artificial island linked to the shore by a 4-mile causeway. Formed with dredged fill material, the site was initially developed for an export coal terminal that opened in 1970. Deltaport construction began early in 1994 and took just over three years to complete. The first containers moved through the new terminal on June 8, 1997, when the American President Line's *President Truman* became the first ship to dock at Deltaport.

The work began with dredging of a ship turning basin and preparations for the sinking of 16 huge precast concrete caissons. These formed a dock face 2198 feet long with a draft of 52 feet at mean low water, large enough to berth two of the largest container ships. The site was excavated to as much as a hundred feet below water level to place select fill material and a rock blanket below the caissons. The 125-foot-long, 60-foot-wide and 70-foot-tall caissons were then sunk into position to provide a foundation for a concrete structure that supports the dock's bollards and fenders and the container-handling gantry cranes.

A 64-acre paved storage yard provides room for the equivalent of 13,000 20-foot containers stacked three high. Four-high platforms provide 420 plug-in points for refrigerated containers, with room for expansion to 600 points. An intermodal rail yard includes a locomotive run-around track and four loading tracks, each 3500 feet long, which provide enough space to load or unload

Two of Deltaport's enormous high-speed gantry cranes unloaded containers from American President Line's container ship *President Kennedy* on September 2, 1997. Each of these Chinese-built cranes stands 238 feet high, can reach out 155 feet over the water, and can load or unload 40 containers an hour.— William D. Middleton.

two 7000-foot-long double-stack intermodal trains at a time. Another 16,000 feet of holding track space is located outside the terminal.

The advanced equipment that makes Deltaport work comes from all over the world. Containers are lifted on and off ships by four of the largest high-speed gantry cranes in North America (a fifth is planned as traffic grows), built at Shanghai, China. Each of these enormous rail-mounted cranes stands 238 feet high and can reach out 155 feet over water, far enough to load and unload a ship 18 containers wide, which is as large as any now afloat or planned. They can lift a maximum load of 55 tons and can load or unload 40 containers an hour. Ten huge British-built, eight-wheel, 45-ton capacity, rubber tired gantry cranes fitted with automatic steering systems move containers between the cranes at the dock face and the storage yard, or on and off trucks.

Two U.S.-built high-speed rail-mounted gantry cranes (another two are planned as needed) serve Deltaport's intermodal rail yard. These have a clear span of 170 feet, enough to straddle all four loading tracks, vehicle access lanes, and container storage areas. They can move through the intermodal yard at 550 feet per minute and can load or unload more than 30 containers an hour.

Linking the different areas of the terminal together is an innovative multi-trailer system that gives Deltaport the handling speed and capacity to unload ships on a "direct hit" basis, with containers moving from ship to rail cars or trucks without ever touching the ground. Each 160-foot unit is made up of a yard tractor pulling a triple-unit trailer rig that can turn on a radius of only 90 feet, and can carry six 20-foot or three 45-foot containers.

This high capacity equipment is tied together by sophisticated computer-based management systems that permit Deltaport to operate on a virtually "paperless" basis. Position determination and equipment identification systems track the location of every container in the terminal electronically and plan and control equipment moves to pick up and load containers. Computerized systems are used for yard and ship planning. Container booking data is provided in advance through an electronic data interchange system, allowing a storage yard slot to be preassigned for every container arriving by either rail or truck. The same system helps to avoid

A fleet of these eight-wheel rubber-tired gantry cranes moves containers between the ship-loading gantries and Deltaport's storage yard and on and off trucks. Here, one of them loads a container onto one of the multi-trailer rigs used to move containers around the terminal.—William D. Middleton.

customs delays by giving Deltaport the ability to pre-clear shipments electronically.

Yard tractors are fitted with cab-mounted lead crystal displays linked to a cargo management program that tell drivers where to position their rigs in the intermodal yard and where to take their loads in the storage yard. Cab data screens tell gantry operators which container cranes they should go to and where to stack or retrieve containers.

All of this, together with time and distance advantages over its rival West Coast ports, has positioned Deltaport and its CN and CP partners to compete for a share of the intermodal traffic that now moves through U.S. West Coast ports. With its high-speed cranes and its capability to move containers directly from ship to rail car, Deltaport can have the first double-stack unit train moving out of the intermodal yard only eight hours after a ship docks, and its unencumbered BC Rail access lets them move directly onto the CP or CN main lines at road speeds.

Both railroads have built on this capability with system-wide improvements that permit double-stack unit train operation all the way to central and eastern Canada or to the U.S. Midwest on schedules competitive with BNSF or UP.

This general view of Deltaport's intermodal yard, where containers are loaded or unloaded from rail cars, shows one of the 170-foot-wide rail-mounted gantries that straddle the entire intermodal yard.—Port of Vancouver.

GETTING THERE

Deltaport is at Roberts Bank, just north of the U.S.-Canadian border and about 25 miles south of downtown Vancouver. The terminal can be reached from Vancouver or Blaine, Washington, via Highway 99 and an access road that leads into the terminal.

The two big rail-mounted gantry cranes that serve Deltaport's intermodal yard can each load or unload 30 containers an hour. One of them loaded containers for an eastbound double-stack train on September 2, 1997.—William D. Middleton.

The Erie's spectacular Cascade Bridge designed by Julius W. Adams was one of the wonders of early railroad engineering. Its 250-foot arch of white oak was said to be the longest single span bridge in the world. This early drawing was by G. N. Todd. —Library of Congress (Neg. LC-USZ62-38597).

5
Lost Landmarks

A remarkable number of the great works of North American railroad engineering have proved extraordinarily durable, many of them continuing to serve well into a second century. But many, too, have fallen victim to storm, fire, flood, or other mishaps, or to simple obsolescence, and it is from among these that we will find our lost landmarks of railroad engineering.

Some, of course, will remain in place almost indefinitely. Barring a few mountain crossings that have been replaced by tunnels, most railroads still follow largely the same route over the mountains laid down by their locating engineers a century and more ago, give or take an occasional line relocation here and there. And once dug or drilled, a tunnel isn't likely to go anywhere; the Allegheny Portage Railroad's Staple Bend Tunnel still stands right where it has been ever since 1832, even though a train hasn't passed through it for 150 years.

Bridges, however, have been a different story. The very early stone viaducts have proved to be among the most enduring engineering works of all. The steel bridges that date from the late nineteenth century and later, too, have demonstrated a remarkable longevity, often with the aid of periodic reinforcing and rebuilding to accommodate the demands of heavier traffic. But the remarkable bridges of wood and iron from the great middle period of railroad development that extended from the 1830s through very nearly the end of the nineteenth century are almost entirely lost to us now and are recalled only through drawings and photographs.

As we have noted, railroad builders turned early to the use of wood for their bridges because they could be erected rapidly and cheaply. The builders threw up these early timber bridges to get their lines into operation as quickly as possible, fully expecting that they would soon have to replace them with more durable structures. Indeed they would, for wooden railroad structures of every kind proved easily susceptible to the ravages of decay or fire or were rendered obsolete by the inexorable advance of locomotive and train weights. The structures of cast and wrought iron that took their place usually lasted but little longer before they, too, required replacement.

The great majority of wooden railroad bridges were simple trestles, or trusses of modest span lengths, but there were some of a truly awesome scale. The Erie's early bridge engineers, while best known for their great stone viaduct at Starrucca, also erected some remarkable wood structures as they built the railroad west across New York. Just a few miles east of Starrucca near Gulf Summit, New York, for example, the railroad spanned the 186-foot-deep chasm of Cascade Creek with the spectacular Cascade Bridge, which was said to be the longest single span bridge in the world. Julius W. Adams, who also designed Starrucca Viaduct, built the bridge with a timber arch of white oak spanning 250 feet. It was a notable attraction while it lasted; the Erie's trains stopped there so passengers could leave the train to admire the bridge and the chasm it spanned. But like so

Near the summit of Sherman Hill in Wyoming the Union Pacific spanned the deep gorge of Dale Creek with this 700-foot timber bridge, which was guyed with ropes and wires against strong winds and the impact of trains.—Union Pacific Railroad Museum (File No. H 2-114).

The Northern Pacific spanned Marent Gulch west of Missoula, Montana, with this massive trestle that was said to be the largest wooden structure in the world. Only two years after it opened the NP replaced it with a steel structure. This drawing of the remarkable span is from *Scientific American* of December 29, 1883.—Library of Congress.

many bridges of wood, it didn't last very long. The bridge opened in 1848. The railroad closed it only seven years later and soon took it down.

The builders of the early railroads to the Pacific, anxious to complete their lines as rapidly as possible, built some massive timber bridges in the mountain ranges of the West in the years following the Civil War. These, too, were usually short-lived. On the western slope of the Black Hills range near the summit of Sherman Hill in Wyoming, the Union Pacific spanned the 126-foot-deep granite gorge of Dale Creek with a 700-foot bridge of double timber bents and Howe trusses, all made from precut mountain, Michigan and Norway pine. Ropes and wires guyed the bridge against the strong Wyoming winds and the impact of trains. Completed in 1868, it was replaced by an iron bridge only eight years later.

An even greater timber trestle was built by the Northern Pacific in 1883 to span the deep Marent Gulch 10 miles west of Missoula, Montana. Over 750 feet long and 226 feet high, it was built with Howe trusses supported by eight timber piers made of timber cut from forests in the vicinity. It was the highest bridge on the Northern Pacific and was said to be the largest wooden structure in the world at the time. NP planned to replace the trestle with one of more durable construction as soon as it could and designed it so that the foundations and piers of a permanent trestle could be erected between those of the timber structure. This came only two years later, when a steel span took the place of the NP's remarkable timber trestle.

The Canadian Pacific was another road that built some expedient timber structures in its rush to the Pacific. Constrained by lack of funds for more permanent construction, as well as anxious to reach Vancouver as quickly as it could, the CP built several extraordinary timber structures for its crossing of the Selkirk Mountains. The Mountain Creek Bridge on the east slope of the Selkirks, for example, was a timber trestle and Howe truss span 164 feet high and 1086 feet long, and it took over two million board feet of lumber to build it. The bridge at Surprise Creek was 180 feet high, while Stoney Creek was spanned by a series of Howe trusses that were higher yet, with a maximum height of more than 228 feet. All were soon replaced by permanent steel spans.

The Attica & Hornellsville's Portage Viaduct (1852) at Portage, New York

The New York & Erie was completed all the way across New York to Dunkirk, on the south shore of Lake Erie, in 1851. This routing missed the important commercial centers developing at Rochester and Buffalo, and new lines were soon under construction that would link the Erie main line with these cities as well.

One of these was the Attica & Hornellsville Railroad, later absorbed by the Erie, which built north and west from a connection with the Erie at Hornellsville (now Hornell) to Attica, where it made connections for Buffalo. The route selected by the line's builders crossed the Genesee River at Portage, New York. They could not have chosen a more difficult place to bridge the stream.

Here, just above the Upper Falls of the Genesee, the river and the parallel Genesee Valley Canal flowed through a ravine some 250 feet deep and nearly 900 feet wide from one bank to the other.

To carry the railroad over this formidable obstacle, Silas Seymour, its superintendent and engineer, designed and built a timber trestle of unprecedented size. Thirteen stone piers were erected in the bed of the river or cut into the banks on either side and were carried to a height of 30 feet to avoid any danger from floods. The stone foundations were surmounted by timber piers 190 feet high. These were built with a batter on each side for stability and were braced to each other by horizontal and diagonal bracing. Timber Howe trusses 50 feet long and 14 feet deep spanned between the piers. The bridge was fabricated so that a defective member could be removed and replaced without disturbing the remainder of the structure.

The immense viaduct stood 234 feet high and spanned 800 feet. Its construction cost $175,000 and required 9200 cubic yards of stone masonry, 108,802 pounds of iron, and some 1.6 million linear feet of timber, which was cut from 246 acres of pine forest. It was said to be the largest wooden bridge in the world at the time of its completion. "This viaduct," said *Engineering News* of it many years later, "was the boldest attempt ever made in timber trestles."

Work began on the giant trestle in the summer of 1850. The railroad had not yet been completed to the site, and materials for its construction were brought in by canal and road. The track finally reached the east side of the ravine from Hornellsville on January 22, 1852, but it was almost seven months more before the viaduct was finally ready for service. On August 9 all was in

Incorporating some 1.6 million linear feet of timber in its construction, the Attica & Hornellsville's Portage Viaduct ranked as the world's largest timber bridge at the time of its completion, and nothing built of wood ever surpassed it. This drawing of the structure is from the *Scientific American* of September 7, 1867.—Author's collection.

Lost Landmarks

As spectacular as the Portage Viaduct itself was the great fire of May 6, 1875, that destroyed it with a roar that one observer likened to "a hurricane approaching through the forest." This drawing of the conflagration appeared in *Frank Leslie's Illustrated Newspaper* for May 22, 1875.—Author's collection.

readiness and the Erie's 4-4-0 *Orange,* which had performed similar duties a few years earlier at the celebrated Starrucca Viaduct, steamed across the trestle with a four-car train to mark its opening. Among the many dignitaries on board for the occasion were New York Governor Washington Hunt and Erie President Benjamin Loder. A formal dedication ceremony on August 25 was followed by a banquet in the bridge workmen's mess hall, which included on its Lucullan menu roast beef from a 3600-pound ox.

With the bridge complete, the new line quickly developed a substantial traffic as Buffalo displaced Dunkirk as the Erie's principal western terminal. The site of the bridge was a splendidly scenic place. Deep in its ravine, the river flowed between rocky walls. Just below the bridge the stream plunged over the 40-foot Upper Falls and then, a quarter mile further along, plunged again over the 80-foot Middle Falls, said to be one of the most beautiful cascades in America. Visitors were drawn to Portage to visit the great wooden bridge and the surrounding scenery, and the Erie ran excursion trains to the site from all over western New York.

Like many another wooden bridge before it the Portage viaduct came to a fiery end. The Erie, mindful of the danger of fire to such a structure, had posted watchmen at the bridge day and night, but to no avail. Around 1 a.m. on May 6, 1875, shortly after the passage of an eastbound train, the watchman discovered a fire in the decking of the bridge. Unable to turn on the fire hose installed on the bridge for just such an eventuality, the watchman ran to give the alarm and left the bridge to its fate. The flames quickly spread to every part of the structure, and the bridge burned for more than three hours. It was a conflagration worthy of this greatest of all timber bridges.

"The spectacle presented at precisely four o'clock was fearfully grand," reported eyewitness W. P. Letchworth in a lurid account for the *Buffalo Courier;* "every timber in the bridge seemed then to be ignited, and an open network of the fire was stretched across the upper end of the valley. . . . The hoarse growl of the flames and the cracking of the timbers sounded like a hurricane approaching through the forest.

"At fifteen minutes past four," wrote Letchworth, "the superstructure of the west end of the bridge sank downward and the depression rolled throughout its length to the east end like the sinking of an ocean wave. The whole upper structure including the heavy T rails, went down with a crashing sound so terrible as it came to our ears on the wind, that it surpassed the prolonged roar of the falling avalanche one may hear at times in spring upon the declivities of the Western Alps."

With what had by now become its main line to Buffalo cut by the disaster, the Erie moved quickly to replace the bridge. Only 47 days later a new Portage viaduct was in place. But this one was built of wrought iron. Silas Seymour's great bridge of wood would never be equaled or surpassed.

The Grand Trunk's Niagara River Suspension Bridge (1855) at Niagara Falls, New York

One of the most remarkable railroad bridges of the nineteenth century was John A. Roebling's great suspension bridge high above the Niagara River. It spanned more than twice as far as any railroad bridge then in existence, and it was the only railroad suspension bridge ever built in North America.

The bridge was built to connect the New York railroads with lines being built across southern Ontario. The great gorge of the Niagara, however, was a formidable barrier to be overcome. At a point below the Falls that seemed the best location for a crossing the gorge was about 800 feet across and over 200 feet deep. This was far too great a span for the railroad bridge building technology of the time; and the turbulent rapids, together with enormous ice jams in the winter, made the construction of piers or falsework impractical.

Only a suspension bridge seemed feasible for a span this long. Several had been built in Europe and the United States, but all of these were road bridges. It was widely believed that a suspension bridge could not be built that was rigid enough to carry railroad traffic, but Roebling succeeded in doing it.

The Niagara Bridge Company was formed in 1846, and Charles Ellet, Jr. won a contract to build the bridge the following year. Ellet soon gave up the work in a dispute with the bridge company, and the company then turned to John Roebling, who had also submitted a proposal for the work. Roebling had already built several suspension bridges, and he was convinced that he could build one rigid enough to carry railroad loadings.

The structure designed by Roebling spanned 821 feet between towers. Two limestone towers topped with Egyptian capitals rose almost 60 feet above the track level at each end of the structure to support the suspension cables. Two 10-inch diameter wire cables on each side of the bridge were spun in place from wrought iron wire using a method developed and patented by Roebling. A continuous wire was drawn from a reel that was carried back and forth across the span by a suspended sheave that left a single wire in the correct position on each passage. Each cable was made up of seven strands, and each strand contained 520 of these wires tightly packed together. At each end of the span the cables were held in place by a cable anchorage carried deep into solid rock.

The bridge decks were suspended from the cables by a series of wire rope suspenders, with the inner cable on each side supporting a single-track rail deck and the outer cables dropping to a lower level to support a roadway. The two decks were carried by a structure of latticed wood and iron trusses 700 feet long, 22 feet wide

—Library of Congress.

John Augustus Roebling (1806–1869) was born in Mühlhausen, Germany, where he was educated in local public schools and privately tutored. He then earned a degree in civil engineering from the Royal Polytechnic Institute in Berlin and took up road building work, at the same time studying bridge building.

Roebling emigrated to the United States when he was 25 to take up farming. This was not a success, and in 1837 he returned to engineering work on canal and railroad projects in Pennsylvania. Roebling became involved with the Allegheny Portage Railroad and conceived the idea of wire rope cables as an improved substitute for the hemp ropes then used on the inclines. In 1841 he successfully manufactured the first wire rope with machinery of his own design. From this beginning Roebling established a great wire rope industrial empire that carried his name.

Roebling also became the great pioneer of the modern suspension bridge. His experience in building two important early suspension spans with wire rope cables led to the commission for the Niagara River bridge, and he later built major suspension bridges at Pittsburgh and Cincinnati. All of these were precursors to Roebling's greatest achievement, the suspension bridge across the East River between Manhattan and Brooklyn. He was appointed its chief engineer in 1867 and designed the greatest bridge of the nineteenth century. Work had begun on final surveys for the bridge in July 1869 when his foot was caught and crushed in an accident at the Fulton Ferry slip. Roebling contracted a tetanus infection and died two weeks later. It fell to his oldest son and civil engineering colleague, Washington Augustus Roebling, to complete the Brooklyn Bridge.

and 18 feet deep. Roebling wanted to assure that the long cable-supported span would have adequate stiffness, and the heavy trusses served this important function in addition to carrying the decks.

Roebling further stiffened the bridge with a system of 64 wire rope cable stays radiating diagonally from the towers to the upper deck of the bridge, and another 56 that extended from the lower deck to the rock walls of the gorge.

Work began on the bridge in September 1851, and it took almost four years to complete it. On March 8, 1855, the first locomotive crossed the completed structure. "No vibrations whatever," Roebling wrote in his

Taken from an early stereopticon view, this photograph of a locomotive on the Niagara suspension bridge shows the four-rail track that was required to accommodate the three different track gauges of the lines that used the span.—Smithsonian Institution.

Depicted here in an early drawing, John Roebling's record suspension bridge across the Niagara River gorge was a tourist attraction almost equal to the river's great falls.—Smithsonian Institution (Neg. 50076).

notebook. Regular service began operating over the bridge about ten days later.

The bridge was a great success, and dozens of trains a day were soon operating over the structure. For 20 years the bridge carried all railroad traffic across the Niagara. But as railroad traffic grew and locomotives and cars became heavier, Roebling's bridge proved to have its limitations. In 1878 the cable anchorages had to be strengthened, and during 1879–80 the timber stiffening trusses were replaced with a stronger iron and steel structure. By 1886 there were problems with the stone towers, and they were replaced with new wrought iron towers. Throughout all of these remedial works the bridge continued to carry traffic.

A decade later the Grand Trunk, which now owned the bridge, decided to replace it. Train loadings had grown to three times those for which the bridge had originally been designed, and trains had to cross the structure at no more than 5 mph. Rail traffic had grown so heavy that a double-track crossing was needed. A new steel arch bridge was erected around the suspension bridge, with traffic continuing across the old span all the while. When the new bridge was complete in July 1897, the Roebling bridge was taken down. The first, and still the only, railroad suspension bridge in North America had become a lost landmark.

This early view of a train on the bridge shows the deep stiffening trusses and the system of wire rope stays below the bridge that Roebling devised to stiffen the structure in high winds.—Smithsonian Institution (Neg. 90-7152).

Lost Landmarks

The Rock Island's Mississippi River Bridge (1856) at Rock Island, Illinois

In 1854 the Chicago & Rock Island became the first railroad to reach the Mississippi River from Chicago. Even before the first train steamed into Rock Island on the Illinois side of the river, work was underway to bridge the Mississippi to Davenport, Iowa, and to build on across Iowa. The bridge would be the first railroad crossing of the Mississippi, and its construction would set in motion a historic legal struggle between steamboat and railroad interests that would only be resolved by the United States Supreme Court.

The alignment chosen for the bridge crossed Rock Island, on the Illinois side of the river, which offered the shortest route across the river and favorable construction conditions. The channel east of the island was only 400 feet wide, while the main navigable channel west of the island was 1400 feet wide, with a depth of 6 to 8 feet at low water.

The superstructure for the 1581-foot main channel crossing was built with arched Howe trusses and was made up of five fixed spans and a draw span at the navigation channel. Each of the 250-foot-long fixed span trusses was built with top and bottom chords and diagonal braces and counter braces built up from heavy timbers. Four heavy timber arches in each span rested on the piers 13 feet below the bottom chords. A system of iron truss rods connected the top and bottom chords and suspended the truss from the arch.

The 286-foot-long swing span draw bridge was supported by a center pier 32 feet in diameter and protected by a 350-foot-long crib that extended up and down the river. The bridge swung on a turntable track to provide a 120-foot clear opening on each side of the pier. The trusses were similar to those in the fixed spans, while a gallows frame over the center of the draw span helped to support the ends of the span when it was in the open position.

Construction of the abutments and piers began in July 1853. These were built of stone and rested on the solid limestone of the river bottom. There were numerous problems and delays, but by the end of 1855 the masonry work was nearing completion and scaffolding for construction of the trusses could be erected on the solid ice of the river. These were completed several months later and on April 21, 1856, the Rock Island's locomotive *Des Moines* crossed over the span into Iowa.

"The last link is now forged in the chain that connects Iowa and the great west with the states of the Atlantic Seaboard," wrote the editor of the *Davenport Gazette*. "The iron band that will span our hemisphere has been welded at Davenport; one mighty barrier has been overcome; the Missouri is yet to be crossed and then the locomotive will speed onward to the Pacific."

Even before it was completed the bridge was caught up in a legal struggle with the steamboat interests that would last for almost a decade. While charters from the two states allowed the railroad to build across the Mississippi to the state line at the center of the navigation channel, navigation interests opposed the construction. The federal government joined them in opposition and in January 1855 intervened to seek an injunction halting the work. Six months later a federal judge at Chicago upheld the rights of the bridge company and the work went ahead.

The next chapter in the struggle came on the morning of May 6, 1856, only two weeks after the bridge had

This old print of the Rock Island's pioneer crossing of the Mississippi shows several of the typical 250-foot arched Howe trusses of timber that made up the span. At the left is the swing span draw bridge for river traffic.—Chicago, Rock Island & Pacific, David P. Morgan Memorial Library.

This heavier timber truss bridge replaced the Rock Island's original Mississippi River bridge in 1866. The photograph probably dates to March 1868, when the structure was damaged by ice.—Chicago, Rock Island & Pacific, David P. Morgan Memorial Library.

opened, with the celebrated wreck of the steamboat *Effie Afton*. After safely passing through the draw span the vessel was about 200 feet above the bridge when one of her side wheels came to a halt, and the boat drifted back against the bridge. A stove was turned over by the impact and the vessel was set on fire. Both the *Effie Afton* and one span of the bridge were destroyed in the resulting conflagration. The owners of the *Effie Afton* filed suit against the bridge company, claiming that the structure was a dangerous obstruction to navigation. When the case came to trial in a federal court at Chicago, the railroad's legal team was headed by a young lawyer from Sangamon County, Illinois. His name was Abraham Lincoln. The future president was no stranger to railroad legal work, and his defense of the railroad's interests was a vigorous one. To prepare for the case, Lincoln visited the bridge site in 1857. In a perhaps apocryphal account of this visit, Lincoln sat on the bridge ties above the stream, and with the help of 12-year-old Benjamin Brayton, whose father had worked on the construction of the bridge, worked out a study of the currents that would help him to argue that the accident could not have happened as the steamboat people alleged.

Lincoln's closing argument in the case, based upon the view that "one man had as good a right to cross a river as another had to sail up or down it," was generally considered to have been convincingly presented. Nevertheless, the jury failed to come to agreement and was discharged. A second trial was scheduled, but the steamboat men later dismissed their action.

This would seem to have been the end of matters, but it wasn't. Early in 1858 there was an attempt in Congress to take some action against the bridge. Nothing came of this, but another suit was filed in May 1858 by a St. Louis steamboat owner who asked that the "bridge be declared a nuisance and ordered removed." A federal judge in Iowa agreed and ordered the removal of the three piers and their superstructure which lay within the State of Iowa. In December 1862 the railroad's appeal finally landed before the U.S. Supreme Court, which reversed the decision.

"According to this assumption," said the Court of the steamboat men's argument, "no lawful bridge could be built across the Mississippi anywhere. Nor could harbors or rivers be improved; nor could the great facilities to commerce, accomplished by the invention of railroads, be made available where great rivers had to be crossed."

The Rock Island bridge could stay right where it was, and the railroads' right to bridge a navigable stream was now firmly established.

Meanwhile, the Rock Island bridge had been quickly repaired after the *Effie Afton* incident but there were soon other problems with the crossing. The increasing weight of locomotives and cars overtaxed the capacity of the trusses, and they had to be strengthened with suspension cables or chains. In 1866, only a decade after it had opened, the Rock Island replaced the truss spans with deeper and heavier ones. This new crossing did not last even as long as the first one; in 1872 the Rock Island moved to a new iron bridge at a different location, and the first railroad crossing of the Mississippi River was removed. Today, only a small fragment of an abutment on the Iowa shore remains from this historic structure.

Lost Landmarks

The Milwaukee & St. Paul's Mississippi River Pontoon Bridge (1874) at Prairie du Chien, Wisconsin

The Milwaukee & Mississippi, one of Wisconsin's earliest railroads, completed a line across the state from Milwaukee to the Mississippi River at Prairie du Chien in 1857. Within the next decade other lines were extended west and northwest into Iowa and Minnesota from McGregor, Iowa, just across the river from Prairie du Chien. By 1867 all of these lines had been acquired by the Milwaukee & St. Paul, completing a route that extended all the way from Milwaukee to the Twin Cities via Prairie du Chien.

This new line was interrupted, however, by the Mississippi at Prairie du Chien. The river here was some 7000 feet wide from bank to bank, made up of navigable channels separated by an island. These were about 2000 feet wide on the Wisconsin side and 1500 feet wide on the Iowa side. Without a bridge it was a slow and sometimes difficult crossing to make.

For almost 20 years freight and passengers were transferred across the river by ferries or by sleighs when the river was frozen. Around 1867 John D. Lawler, who had contracted with the M&StP to provide this transportation, began loading freight cars directly onto barges or car floats from tracks on the banks of the river and towing them across with steamboats. During the winter he cut a trench through the ice from shore to shore for the car floats.

This arrangement didn't work very well, and Lawler then built a pile trestle across the river, leaving openings in each channel for steamboats to pass through. Cars were then ferried by car float across these shorter openings, which were filled in with temporary trestles every winter when navigation was suspended. This arrangement was an improvement, but the crossing was still costly and time-consuming. Sometimes, particularly in windy weather, it was a treacherous process as well, with an occasional car lost over the side of the car floats.

In 1872, determined to find a better way of making the transfer, Lawler hired Michael Spettel, one of five brothers from a German shipbuilding family, and set him up in a shop at Prairie du Chien. There, Spettel worked out his innovative ideas for a pontoon draw bridge. It was not an entirely new idea. In 1851 Henry R. Campbell, the chief engineer of a New York railroad, had built a pontoon draw span on Lake Champlain near Rouses Point. This used a 300-foot pontoon that was connected to trestle approaches at each end with adjustable aprons and could be pulled out of the way for water traffic.

Spettel developed a much improved version of this idea with huge pontoons over which trains operated on a raised track. Normally, the pontoons spanned the navigation channel, but they could be floated out of the way to permit river traffic to pass through. A key feature of Spettel's design was its provision for handling different river levels. The water level at Prairie du Chien could vary as much as 22 feet between high and low water, and the elevated structure that supported the track was regulated to the different stages of the river by a system of blocking that was confined in a frame and adjusted by hydraulic jacks. A 30-foot apron of iron and timber trusses at each end, spanning between the trestle approach and the pontoon, connected the tracks on the trestle and the pontoon and allowed for any difference in elevation.

Early in 1874 Lawler constructed a new bridge incorporating Spettel's pontoon. The overall length of the single-track structure was about 8000 feet, crossing both channels and the island with a timber pile trestle, with pontoon draw spans filling the navigation openings. Each of these was a single float 30 feet wide at the bottom and 41 feet wide on the deck, 6 feet deep, and 408 feet long. At one end the pontoon draw spans were attached to a fixed piling by a long pivot arm, allowing the pontoon to swing through an arc of 90 degrees to an open position parallel to the flow of the river. Each pontoon was pulled to the open or closed position by a chain powered from a steam engine. A simple coupling device that could be operated by a single man was used to lock the draw span in the closed position.

The structure was completed at about one-sixth of the cost of a conventional draw span, according to Lawler; and the first train crossed the new span on April

A stereopticon view of the Prairie du Chien bridge, dating to about 1880-1890, shows the link between the floating span and the adjacent timber trestle.—H. R. Farr Photo, State Historical Society of Wisconsin (Neg. WHi (X313)2728).

A view of the west span of the bridge from Marquette, Iowa, shows the pontoon structure in the closed position.—Milwaukee Road.

From the vantage point of the adjacent highway suspension bridge, a photographer recorded the passage of one of the Milwaukee Road's F-3 Class 4-6-2 Pacifics across the pontoon bridge with an eastbound train. The locomotive was about to enter the 276-foot west pontoon span.—State Historical Society of Wisconsin (Neg. WHi (X3)37595).

15, 1874. The arrangement was an immediate success, providing a capacity to move as many as 1000 cars a day across the river. The structure went on to many years of service for the Milwaukee Road.

The bridge was rebuilt twice over the next 40 years, and during 1914 and 1916 major changes were made to the span in yet another rebuilding. Newer pontoons of steel and wood construction were built, and portions of the draw spans were filled in with fixed steel truss spans. But even as rebuilt, the structure retained the basic features of Spettel's original design.

The Milwaukee Road seemed to like the idea of a pontoon draw bridge and later built two other similar bridges. The first of these was built in 1882 for a crossing of the Mississippi at Reeds Landing, near Wabasha, Minnesota, for the Milwaukee's Chippewa Valley branch. Another was completed 1905 for a crossing of the Missouri River at Chamberlain, South Dakota, for the railroad's Rapid City line.

All three of the Milwaukee Road's innovative pontoon draw spans served the railroad long and well. The bridge at Chamberlain was replaced in 1919 by another

Lost Landmarks

pontoon bridge, which remained in service until 1953, when it was replaced by a conventional bridge. The Reeds Landing bridge continued in service until it was damaged by high water in 1951, and it was abandoned the following year.

John Lawler's Prairie du Chien pontoon bridge lasted the longest of all. After passenger service across the bridge ended in January 1960, the bridge carried only three local freights a week and the Milwaukee petitioned the Interstate Commerce Commission to abandon the structure. This approval came in September 1961 and the last train crossed the bridge on October 31, 1961. The west span was moved up the river to LaCrosse the following month and dismantled in 1964. The east span was removed in 1962, leaving few traces of this most innovative of railroad bridges.

The Florida East Coast's Florida Keys Bridges (1912)

Surely one of the most daring schemes from the era of railroading's great builders was Henry M. Flagler's railroad across the sea along the islands of the Florida Keys to Key West. Flagler was a multimillionaire who began developing Florida resorts and then got into the railroad business. By 1886 his Florida East Coast extended all the way to Miami.

A principal reason for the railroad's extension to south Florida was the idea of a deep water port that would link the FEC with trade routes to Cuba, the West Indies, and South America. Miami proved ill suited for a major port, and Flagler was soon looking further south for a better location. By 1904 the daring idea of a line to Key West had come to the fore. Construction of the Panama Canal had begun, and a deep water port on the Florida Keys would place the FEC far closer to the Canal than any other U.S. railroad. Henry Flagler was soon committed to the vision of a railroad over the sea, and construction was underway before the end of 1905.

It was an audacious undertaking. Only 22 of the 128 miles between the end of track at Homestead and Key West were on the mainland. The remainder was built across the chain of keys or small islands that separated the Atlantic Ocean from the Gulf of Mexico. About 30 miles of water crossing were built through a region in which savage hurricanes were common.

Between the mainland and Knights Key there were five major bridges. The longest of these was at Long Key, where the railroad built a 2-mile-long reinforced concrete bridge made up of 180 semicircular arches spanning 50 feet and supported on concrete piers that were founded on solid rock or piling.

Just below Knights Key the builders encountered the longest stretch of open water on the entire extension, where 7 miles of water anywhere from 18 to 22 feet deep had to be bridged. This was accomplished with three distinct structures. A five-mile section of this crossing was made up of 335 steel plate girder spans 60 or 80 feet long that were carried by concrete piers supported on piling driven into the limestone bedrock. At Moser Channel, the connecting link through the Keys between the Atlantic and the Gulf, a 253-foot steel truss swing span was erected; while at the southern end of the Knights Key crossing, a long viaduct of 210 reinforced concrete arches, each with a 35-foot span, was constructed over shallower water.

Further south at Bahia Honda, where water depths of 20 to 30 feet were found, the railroad built a 5100-foot bridge made up of steel trusses and plate girders. There were 26 trusses spanning 128 or 186 feet and one of 247 feet and nine 80-foot plate girder spans. The southernmost of the extension's major crossings was a 2573-foot bridge at Boca Chica, just north of Key West, that comprised 83 reinforced concrete arches, each spanning 25 feet.

At Key West itself there was no land available for a terminal. The railroad created its own by pumping in

At Long Key the Florida East Coast built a 2-mile bridge made up of 180 semicircular reinforced concrete arches each spanning 50 feet. This construction view shows the erection of wooden forms for the concrete arches.—Library of Congress (Neg. HAER FLA 44-KNIKE 1-52).

Completion of the Long Key bridge enabled the FEC to begin passenger train service as far south as Knights Key, where they connected with steamers of the Peninsular & Occidental Steamship Company for Havana. The first train to make the trip was this three-car *Flagler Special,* shown on the Long Key bridge on February 5, 1908.—Library of Congress.

thousands of cubic feet of dredged material to fill a 138-acre site and built a 1700-foot reinforced concrete pier.

The construction of the extension was a massive undertaking for the railroad. Constructing engineer Joseph C. Meredith headed a work force that eventually ranged between 3000 and 4000 men. The enormous fleet of equipment assembled for the work included 30 launches, six stern-wheel Mississippi River steamboats, three tugs, three floating pile drivers, one floating machine shop, and more than a hundred barges and lighters. Eight complete work boats were used, each made up of a barge equipped with a concrete mixer and engine, a boiler, derricks fitted with booms and orange peel buckets, and hoppers for sand and stone supplies. There were 14 two-story houseboats, each capable of accommodating more than a hundred men, to house the workers.

Virtually all materials had to be brought in from the mainland. Concrete materials were shipped in by steamer from as far away as New York, while a special Portland cement used for all concrete below high tide came from Germany by tramp steamers. All fresh water had to be brought from the mainland.

By the time the line was finished it had required 19,000 tons of structural steel, 2000 tons of reinforcing steel, and 800,000 barrels of cement. Some 96,000 tons of rock and another 78,000 tons of gravel were shipped in, while 300,000 cubic yards of coral were dredged from the sea bottom. A total of 461,000 cubic yards of concrete was placed, and 20 million cubic yards of rock, sand and marl were used in the embankments.

The work began in April 1905. There were some extraordinary challenges over the next seven years, and hurricanes proved by far the worst of them. The most damaging blow came in 1906, when construction equipment and already completed work were severely damaged. Many of the laborers had taken refuge in their quarters on the houseboats, some of which were blown out to sea. At least 130 men were lost, and many others were picked up by passing steamers days later. After this, permanent camps for the workers were established on land.

Before the project was complete, two more hurricanes struck in 1909 and 1910 that were the most severe yet seen in coastal Florida. Wind velocities of 125 mph were recorded in the 1909 storm. The builders were better prepared this time, and relatively little loss of life or damage was caused by either storm. Upon warning of a storm the equipment on land was protected, and floating equipment was taken to shallow water and sunk. It was considered better to have to raise the equipment and repair it than to risk even worse damage from a hurricane. The worst damage suffered in the 1909 storm was the loss of five plate girders that were blown off their piers on the Knights Key bridge.

A different and unanticipated problem revealed by the 1909 and 1910 storms was an impounding of the waters of the Gulf by hurricane winds in the Bay of Florida, which lay between the Keys and the Florida coast. This was attributed to the closing off of much of the drainage through the Keys by the railroad's embankments. Held in the Bay by winds, the water built up until it overtopped and washed out the embankments. The problem was solved only by replacing miles of embankment with concrete arch spans and by adding protection to other sections of the embankment.

Anxious to see the line complete while an aging and ill Henry Flagler still lived, his associates in February 1911 ordered the work completed by January 1912, a year earlier than had been planned. Constructing engineer William J. Krome, who had succeeded to the job upon Meredith's death in 1909, managed to do it. Lights were rigged at the construction sites so that work could proceed around the clock. Portions of the line were completed only in a temporary fashion, with more permanent construction to come later.

The final plate girder was lifted into place on the Knights Key bridge on the morning of January 21, 1912, and the last spike for the Key West Extension was driven

At Key West, FEC trains offered passengers a dockside transfer to steamers for Havana, Cuba. Pullman passengers could travel all the way from New York to Key West aboard the *Havana Special* without changing cars.—Library of Congress.

the following day. Later in the day a five-car *Extension Special* departed from Miami for Key West. Aboard it in his private car *Rambler* was Henry Flagler. On the train's arrival in Key West it was greeted by a crowd estimated at 10,000. Bands played, ships in the harbor blew their whistles, and the city began a three-day fiesta.

Service over the extension included the new *Overseas Limited*, which carried through Pullmans between New York and Key West, where passengers made a dockside transfer to or from the steamships which served Havana. Florida East Coast advertising promoted the unique experience of "Going to Sea by Rail." Over the next few years the railroad built three big car ferries to carry freight cars to and from the Cuban railroad system.

But the costly extension was never the success that Flagler had hoped for. Through passenger service was limited to the daily *Havana Special*. A modest freight traffic with Cuba developed, but the anticipated traffic with Central and South America or the Panama Canal never materialized, and a great terminal Flagler had planned at Key West was never built.

The Key West Extension was never able to pay off the FEC's enormous investment, or even to meet the high costs of maintaining the line. The railroad entered re-

Florida East Coast advertised the experience of "Going to Sea by Rail," and this classic 1912 view of the *Havana Special* on the 2-mile Long Key bridge suggests what the journey was like.—Florida East Coast, David P. Morgan Memorial Library.

Henry Flagler's great dream ended abruptly in the fierce hurricane of September 1935. This was the fate of a rescue train dispatched to rescue highway workers at Islamorada. The 11-car train was swept off the rails by a 17-foot tidal wave; only the locomotive and tender remained on the rails.—Charles Layng, David P. Morgan Memorial Library.

ceivership in 1931 and was ill equipped to deal with the blow to the fortunes of the line that came on September 2, 1935.

On that day, a fearful hurricane swept across the Keys. A train dispatched to Islamorada to rescue highway workers was swept off the rails by a 17-foot tidal wave, and hundreds lost their lives. The railroad's bridges held firm, but much of the roadbed was swept away over the 42 miles between Key Largo and Key Vaca. Facing repair expenses estimated at anywhere from two to three million dollars, the bankrupt railroad decided to abandon the extension instead. Henry Flagler's railroad over the sea had lasted just a little more than 23 years.

The Florida East Coast sold what remained of the line to the State of Florida. The right-of-way and the bridges provided a route for a new Overseas Highway that opened to Key West in 1938. Although it hasn't seen a train since 1935, Henry M. Flagler's not-quite-lost landmark of railroad engineering continues to provide a useful transportation service to this day.

Lost Landmarks

Bibliography

Across the Waters
Allen, Richard Sanders, "Crossings under Cover," *Trains*, Vol. 15, No. 8 (June 1955): 44–50.
American Wooden Bridges. New York: American Society of Civil Engineers, 1976.
Cook, Richard J., *The Beauty of Railroad Bridges*. San Marino, Calif.: Golden West Books, 1987.
DeLony, Eric, *Landmark American Bridges*. Boston, Mass.: Bullfinch Press, 1993.
Gross, H. H., "Steel across the Rivers," *Railroad Magazine*, Vol. 44, No. 4 (January 1948): 8–38, and Vol. 45, No. 1 (February 1948): 10–27.
Jackson, Donald C., *Great American Bridges and Dams*. Washington, D.C.: Preservation Press, 1988.
Plowden, David, *Bridges: The Spans of North America*. New York: W. W. Norton, 1974.

CARROLLTON VIADUCT AND THOMAS VIADUCT
Dilts, James D., *The Great Road*. Stanford, Calif.: Stanford University Press, 1993.
Harwood, Herbert H., Jr., *Impossible Challenge II*. Baltimore, Md.: Barnard, Roberts, 1994.

CANTON VIADUCT
Galvin, Edward D., *A History of Canton Junction*. Brunswick, Maine: Sculpin Publications, 1987.
———. "The Canton Viaduct," *Railroad History*, No. 129 (Autumn 1973): 71–85.

STARRUCCA VIADUCT
Young, William S., *Starrucca: The Bridge of Stone*. Privately published, 1995.

THE BOLLMAN TRUSS BRIDGE
Dilts, James D., *The Great Road*. Stanford, Calif.: Stanford University Press, 1993.
Harwood, Herbert H., Jr., *Impossible Challenge II*. Baltimore, Md.: Barnard, Roberts, 1994.
Vogel, Robert M., "The Engineering Contributions of Wendel Bollman," Paper 36, *Bulletin 240: Contributions from the Museum of History and Technology*. Smithsonian Institution (1964): 79–97.

VICTORIA BRIDGE
"The Victoria Bridge," *Bulletin No. 23*, Railway & Locomotive Historical Society (1930): 57–68.
Triggs, Stanley; Brian Young; Conrad Graham; and Gilles Lauzon, *Victoria Bridge—The Vital Link*. Montreal: McCord Museum of Canadian History, 1992.

THE ST. LOUIS BRIDGE
Diaz, David, "Under Pressure," *American Heritage of Invention & Technology*, Vol. 11, No. 4 (Spring 1996): 52–63.
Gies, Joseph, "Mr. Eads Spans the Mississippi," *American Heritage*, Vol. 20, No. 5 (August 1969): 16–21, 89–93.
Kouwenhoven, John A., "The Designing of the Eads Bridge," *Technology and Culture*, Vol. 33, No. 4 (October 1982): 535–568.
Scott, Quinta, *The Eads Bridge*. Columbia: University of Missouri Press, 1979.

KENTUCKY RIVER BRIDGE
Curry, Howard, *High Bridge: A Pictorial History*. Lexington, Ky.: Howard Curry, 1984.

MINNEAPOLIS STONE ARCH BRIDGE
Historic Stone Arch Bridge: The Reopening of a Landmark. Minneapolis, Minn.: St. Anthony Falls Heritage Board, 1994.
"Stone Arch Bridge," *Northstar News*, Northstar Chapter, National Railway Historical Society, Minneapolis–St. Paul, Minnesota (May 1972): 4–5.

POUGHKEEPSIE BRIDGE
Hall, Ronald J., "The Poughkeepsie High Bridge," *Shoreliner*, Vol. 10, Issue 4 (1979): 5–11.
McLaughlin, D. W., "Poughkeepsie Gateway," *Bulletin No. 119*, Railway & Locomotive Historical Society (October 1968): 6–33.

LUCIN CUT-OFF
"Biggest Job since Promontory," *Trains*, Vol. 17, No. 3 (January 1957): 9.
Mann, David H., "Bridging Great Salt Lake," *Railroad*, Vol. 36, No. 1 (June 1944): 8–25.

TUNKHANNOCK VIADUCT
Taber, Thomas Townsend and Thomas Townsend Taber III, *The Delaware, Lackawanna & Western Railroad in the Twentieth Century: 1899–1960*. Muncy, Pa.: Thomas T. Taber III, 1980.

HELL GATE BRIDGE
Buckley, Tom, "The Eighth Bridge," *The New Yorker*, Vol. 66, No. 48 (January 14, 1991): 37–59.
Middleton, William D., *Manhattan Gateway*. Waukesha, Wis.: Kalmbach Books, 1996.

QUEBEC BRIDGE
Tarkov, John, "A Disaster in the Making," *American Heritage of Invention & Technology,* Vol. 1, No. 3 (Spring 1986): 10–17.
Worthen, Sanborn S., "The Bridge That Balked," *Trains,* Vol. 45, No. 5 (March 1985): 26–32.

METROPOLIS BRIDGE
Mitchell, John D., Jr., "The Q in the Coal Fields," Zeigler, Ill.: Burlington Route Historical Society, 1999.
Overton, Richard C., *Burlington Route,* New York: Alfred A. Knopf, 1965.

HUEY P. LONG BRIDGE
Masters, Frank M., *Mississippi River Bridge at New Orleans, Louisiana: Final Report.* Harrisburg, Pa.: Modjeski and Masters, Engineers, 1941.

Across Great Mountains

Allen, C. Frank, *Railroad Curves and Earthwork.* New York: McGraw-Hill, 1931.
Budd, Ralph, "The Conquest of the Rockies," *Trains,* Vol. 7, No. 12 (October 1947): 42–55.
Crandall, Charles Lee and Fred Asa Barnes, *Railroad Construction.* New York: McGraw-Hill, 1913.
Modelski, Andrew M., *Railroad Maps of North America: The First Hundred Years.* Washington, D.C.: Library of Congress, 1984.
Sprague, Marshall, *The Great Gates: The Story of the Rocky Mountain Passes.* Boston, Mass.: Little, Brown, 1964.

THE BERKSHIRE HILLS CROSSING
Fisher, Charles E., "Whistler's Railroad," *Bulletin No. 69,* Railway & Locomotive Historical Society (1947): 8–100.
Alvin F. Harlow, "George Washington Whistler," *Trains,* Vol. 8, No. 6 (April 1948): 14–18.
———, *Steelways of New England.* New York: Creative Age Press, 1946.

THE ALLEGHENY MOUNTAIN CROSSING
Morgan, David P., "World's Busiest Mountain Railroad," *Trains,* Vol. 17, No. 6 (April 1957): 16–30.
Roberts, Charles S., assisted by Gary W. Schlerf, *Triumph I: Altoona to Pitcairn, 1846–1996.* Baltimore, Md.: Barnard, Roberts, 1997.
Sohlberg, Harry T., "Horseshoe Curve," *Trains,* Vol. 1, No. 5 (March 1941): 28–35.

DONNER PASS
Hinckley, Helen, *Rails from the West: A Biography of Theodore D. Judah.* San Marino, Calif.: Golden West Books, 1969.
Kraus, George, *High Road to Promontory.* Palo Alto, Calif.: American West, 1969.
Signor, John R., *Donner Pass: Southern Pacific's Sierra Crossing.* San Marino, Calif.: Golden West Books, 1985.

MALTRATA INCLINE
DeGolyer, Everett L., Jr. and Stan Kistler, "Mexicano!" *Trains,* Vol. 21, No. 7 (May 1961): 15–25.

TEHACHAPI PASS
Houghton, Richard, "Over the Tehachapi," *Trains,* Vol. 7, No. 2 (December 1946): 14–21.
Morgan, David P., "Earthquake!" *Trains & Travel,* Vol. 13, No. 1 (November 1952): 14–20.
Signor, John R., *Tehachapi.* San Marino, Calif.: Golden West Books, 1983.

ROGERS PASS
Berton, Pierre, *The Impossible Railway.* New York: Alfred A. Knopf, 1970.
Lavallée, Omer, *Van Horne's Road.* Montreal: Railfare Enterprises, 1974.

MARIAS PASS
Brimlow, George F., ed., "Marias Pass Explorer, John F. Stevens," *Montana Magazine of History,* Vol. 3, No. 3 (Summer 1953): 39–44.
Coy, John R. and Robert C. Del Grosso, *Montana's Marias Pass: Early GN Mileposts and BNSF Guide.* Bonners Ferry, Idaho: Great Northern Pacific Publications, 1996.
Martin, Albro, *James J. Hill and the Opening of the Northwest.* New York: Oxford University Press, 1976.
Wood, Charles and Dorothy, *The Great Northern Railway: A Pictorial Study.* Edmonds, Wash.: Pacific Fast Mail, 1979.

THE CROSSING OF THE SIERRA MADRE OCCIDENTAL
Kerr, John Leeds with Frank P. Donovan, *Destination Topolobampo.* San Marino, Calif.: Golden West Books, 1968.

Railroads below Ground

Beaver, Patrick, *A History of Tunnels.* Secaucus, N.J.: Citadel Press, 1972.
Black, Archibald, *The Story of Tunnels.* New York: Whittlesey House, 1937.
Brunton, David W., and John A. Davis, *Modern Tunneling.* New York: John Wiley & Sons, 1914.
Covington, Stuart, "Railroad Tunnels," *Trains,* Vol. 6, No. 7 (May 1946): 11–21.
Lauchli, Eugene, *Tunneling.* New York: McGraw-Hill, 1915.
Prelini, Charles, with additions by Charles S. Hill, *Tunneling: A Practical Treatise.* New York: D. Van Nostrand, 1901.
Richardson, Harold W., and Robert S. Mayo, *Practical Tunnel Driving.* New York: McGraw-Hill, 1941.
Vogel, Robert M., "Tunnel Engineering—A Museum Treatment." Paper 41, *Bulletin 240: Contributions from the Museum of History and Technology.* Smithsonian Institution (1964): 203-239.

BLUE RIDGE TUNNEL
Hunter, Robert F. and Edwin L. Dooley, Jr., *Claudius Crozet, French Engineer in America, 1790–1864.* Charlottesville: University Press of Virginia, 1989.
Warden, William E., Jr., "Claudius Crozet, Napoleon's Captain versus the Blue Ridge," *Railroad History,* No. 129 (Autumn 1973): 44–55.

HOOSAC TUNNEL

Allen, Richard Sanders, "The Great Bore," *Trains*, Vol. 20, No. 8 (June 1960): 18–26.

Byron, Carl R., *A Pinprick of Light: The Troy and Greenfield Railroad and Its Hoosac Tunnel*. Shelburne, Vt.: New England Press, 1995.

Meyer, William B., "The Long Agony of the Great Bore," *American Heritage of Invention & Technology*, Vol. 1, No. 2 (Fall 1985): 52–57.

ST. CLAIR RIVER TUNNELS

Pinkepank, Jerry A., "A Tale of Two Tunnels," *Trains*, Vol. 24, No. 11 (September 1964): 36–44; and Vol. 24, No. 12 (October 1964): 40–47.

"The Iron Link: Port Huron–Sarnia Railway Tunnel," in *Michigan History*, Vol. 54, No. 1 (Spring 1970): 62–72.

B&O HOWARD STREET TUNNEL

Duke, Donald, "America's First Main Line Electrification," *American Railroad Journal*—1966: 86–109.

CPR SPIRAL TUNNELS

Berton, Pierre, *The Impossible Railway*. New York: Alfred A. Knopf, 1970.

Lavallée, Omer, *Van Horne's Road*. Montreal: Railfare Enterprises, 1974.

Pole, Graeme, *The Spiral Tunnels and the Big Hill*. Canmore, Alberta: Altitude Publishing Canada, 1995.

PRR HUDSON RIVER TUNNELS

Condit, Carl W., *The Port of New York: A History of the Rail and Terminal System from the Beginnings to Pennsylvania Station*. Chicago: University of Chicago Press, 1980.

Middleton, William D., *Manhattan Gateway*. Waukesha, Wis.: Kalmbach Books, 1996.

DETROIT RIVER TUNNEL

Pinkepank, Jerry A., "A Tale of Two Tunnels," *Trains*, Vol. 24, No. 11 (September 1964): 36–44, and Vol. 24, No. 12 (October 1964): 40–47.

MOFFAT TUNNEL

Albi, Charles and Kenton Forrest, *The Moffat Tunnel: A Brief History*. Golden, Colo.: Colorado Railroad Museum, 1978.

Boner, Harold A., *The Giant's Ladder*. Milwaukee, Wis.: Kalmbach Publishing, 1962.

Crossen, Forest, "David Moffat's Dreams Come True," *Trains*, Vol. 2, No. 3 (January 1942): 28–36.

CASCADE TUNNEL

Hilton, Jerrold F., "The 50th Anniversary of the Cascade Tunnel," *National Railway Bulletin*, Vol. 44, No. 4 (1979): 4–16, 46.

McLaughlin, D. W., "Cascade Passage," *Trains*, Vol. 22, No. 1 (November 1961): 22–29; Vol. 22, No. 2 (December 1961): 22–33.

White, Victor H., "Through Cascade Tunnel," *Trains*, Vol. 1, No. 8 (June 1941): 4–9.

MOUNT MACDONALD TUNNEL

Keefe, Kevin P., "Third Conquest of Rogers Pass," *Trains*, Vol. 50, No. 5 (March 1990): 30–41.

Yards, Docks and Terminals

DeBoer, David J., *Piggyback and Containers: The History of Rail Intermodal on America's Steel Highway*. San Marino, Calif.: Golden West Books, 1992.

Droege, John A., *Freight Terminals and Trains*. New York: McGraw-Hill, 1925. Reprint edition, Chattanooga, Tenn.: National Model Railroad Association, 1998.

Josserand, Peter, "Push-Button Yards," *Railroad*, Vol. 69, No. 6 (October 1958): 20–27.

Shaffer, Frank E., "Rundown on Automation," *Trains*, Vol. 21, No. 5 (March 1961): 22–23.

Strawbridge, Kenneth, "The Landships," *Trains*, Vol 54, No. 3 (March 1994): 58–64.

White, John H., Jr., *The American Railroad Freight Car*. Baltimore, Md.: Johns Hopkins University Press, 1993.

Williams, Jay, "Ore-Dock Dinosaurs," *Trains*, Vol. 51, No. 4 (February 1991): 50–55.

B&LE DOCK NO. 4

Beaver, Roy C., *The Bessemer and Lake Erie Railroad, 1869–1969*. San Marino, Calif.: Golden West Books, 1969.

ENOLA YARD

Comstock, Henry B., "Redball Terminal," *Railroad Magazine*, Vol. 30, No. 6 (November 1941): 6–25.

Cupper, Dan and Robert S. McGonigal, "Enola Yard Closes: Who's Next?" *Trains*, Vol. 54, No. 1 (January 1994): 14–15.

Hastings, Philip R., "12,000 Cars a Day," *Trains & Travel*, Vol. 13, No. 11 (September 1953): 22–27.

DULUTH ORE DOCK NO. 6

Dorin, Patrick C., *The Lake Superior Iron Ore Railroads*. Seattle, Wash.: Superior Publishing, 1969.

King, Frank A., *Missabe Road . . . The Duluth, Missabe & Iron Range Railway*. San Marino, Calif.: Golden West Books, 1972.

PROVISO YARD

"C&NW's Proviso Yard," *Trains*, Vol. 4, No. 12 (October 1944): 6–9.

Grant, H. Roger, *The North Western: A History of the Chicago & North Western Railway System*. De Kalb, Ill.: Northern Illinois University Press, 1996.

"Proviso Yard—1944" (repr. from October 1944 *Trains*), "Proviso Yard—Yesterday," by Robert A. Janz, and "Proviso Yard—Today," by Joe Piersen, in *North Western Lines*, Vol. 16, No. 1 (Winter 1989): 16–33.

NORFOLK COAL PIER NO. 6

Striplin, E. F. Pat, *The Norfolk & Western: A History*. Roanoke, Va.: Norfolk & Western Railway, 1981.

Lost Landmarks

THE PORTAGE VIADUCT

Anderson, Mildred Lee Hills, *Genesee Echoes*. Dansville, N.Y.: F. A. Owen, 1956.

Hungerford, Edward, *Men of Erie*. New York: Random House, 1946.

Storey, Edwin C., "The Erie in the Genesee Valley," *Railway Bulletin*, Vol. 1, No. 6 (1976): 4–10.

THE NIAGARA RIVER SUSPENSION BRIDGE

Allen, Richard Sanders, "Over Niagara on a Wire," *Trains*, Vol. 18, No. 2 (December 1957): 44–47.

Spanning Niagara: The International Bridges 1848–1962. Seattle: University of Washington Press, 1984.

Steinman, D. B., *The Builders of the Bridge: The Story of John Roebling and His Son.* New York: Harcourt, Brace, 1945.

THE ROCK ISLAND BRIDGE

Agnew, Dwight L., "Jefferson Davis and the Rock Island Bridge," *Iowa Journal of History*, Vol. 47, No. 1 (January 1949): 3–14.

Beatty, Albert R., "Lincoln and the Railroads," *Trains*, Vol. 5, No. 4 (February 1946): 40–46.

Fowle, Frank P., "The Original Rock Island Bridge Across The Mississippi River," *Bulletin No. 56*, Railway & Locomotive Historical Society (1941): 55–63.

Hayes, William Edward, *Iron Road to Empire.* New York: Simmons-Boardman, 1953.

Starr, John W., Jr., *Lincoln and the Railroads.* New York: Dodd, Mead, 1927.

Zobrist, Benedict K., "Steamboat Men versus Railroad Men: The First Bridging of the Mississippi River," *Missouri Historical Review*, Vol. 59, No. 2 (January 1965): 159–172.

THE PRAIRIE DU CHIEN PONTOON BRIDGE

Hegeman, Jeanette, "The Bridge That Floats," *Trains & Travel*, Vol. 12, No. 3 (January 1952): 43–45.

Miller, Alden E., "The Prairie du Chien Pontoon Bridge," *Bulletin No. 58*, Railway & Locomotive Historical Society (1942): 46–54.

THE FLORIDA KEYS BRIDGES

Layng, Charles, "The Railroad over the Sea," *Trains*, Vol. 8, No. 1 (November 1947): 44–48.

Morgan, David P., "He Dared the Ocean," *Trains & Travel*, Vol. 13, No. 3 (January 1953): 48–49.

Parks, Pat, *The Railroad That Died at Sea.* Key West, Fla.: Langley Press, 1990.

General References

Condit, Carl W., *American Building Art* (2 volumes). New York: Oxford University Press, 1960, 1961.

Hay, William W., *Railroad Engineering.* New York: John Wiley & Sons, 1982.

Jacobs, David, and Anthony E. Neville, *Bridges, Canals & Tunnels.* New York: American Heritage, 1968.

Raymond, William G., Henry E. Riggs, and Walter C. Sadler, *Elements of Railroad Engineering.* New York: John Wiley & Sons, 1941.

Schodek, Daniel L., *Landmarks in American Civil Engineering.* Cambridge: Massachusetts Institute of Technology, 1987.

The American Railway. New York: Charles Scribner's Sons, 1892.

Vance, James E., Jr., *The North American Railroad.* Baltimore, Md.: Johns Hopkins University Press, 1995.

Ward, James A., *That Man Haupt: A Biography of Herman Haupt.* Baton Rouge, La.: Louisiana State University Press, 1973.

Webb, Walter Loring, *Railroad Construction: Theory and Practice.* New York: John Wiley & Sons, 1922.

Historic American Engineering Record

The research reports of the Historic American Engineering Record (HAER) program administered by the National Park Service have been a valuable reference resource for a number of the structures included in this volume. These reports are on deposit and available at the Prints and Photographs Division of the Library of Congress.

Other Periodicals

Descriptions of the design and construction of many of the engineering works included in this volume are based upon the detailed accounts that were published in the various technical and professional journals of the time. Chief among the railroad trade journals have been *Railway Age* and its predecessors *The Railroad Gazette, Railway Age Gazette,* and *American Railroad Journal and Mechanic's Magazine;* and *Railway and Marine News.* Principal engineering and construction journal sources have included *Compressed Air Magazine* and *Engineering News-Record* and its predecessors *The Engineering Record, Engineering News* and *Engineering News and American Contract Journal.* Professional journals consulted have included the *Journal of the Western Society of Engineers, Transactions of the American Society of Civil Engineers,* and the ASCE's magazine *Civil Engineering.*

Other valuable periodical sources have included *General Electric Review, Scientific American,* and *Scientific American Supplement,* which covered technological and scientific developments of the latter part of the nineteenth century and early twentieth century in remarkable detail; and a wide range of local newspapers.

Biographical

Principal sources for the biographical profiles of engineers include *A Biographical Dictionary of American Civil Engineers,* published in two volumes by the American Society of Civil Engineers in 1972 and 1991; the *Dictionary of American Biography;* the *Dictionary of Canadian Biography;* memoirs published in *Transactions of the American Society of Civil Engineers; The National Cyclopedia of American Biography;* the series "People in Public Works," published in the American Public Works Association *Reporter;* and profiles and obituaries published in *Canadian Consulting Engineer, Engineering News, The Illustrated London News, Military History,* and various newspapers.

Index

Adams, Charles Francis, 78
Adams, Julius Walker, 18–19, 170–171; bio., 19
Allegheny Portage RR, 103–104, 171, 175
Allen, C. Frank, 69
Ambassador Bridge (Mich.), 52
American Bridge Co., 6, 63
American Institute of Steel Construction (A.I.S.C.), 65–66
American Railway Engineering Assn. (A.R.E.A.), 9
American Society of Civil Engineers (A.S.C.E.), 11, 19, 39, 97, 128
    International Historical Civil Engineering Landmarks, 53
    National Historic Civil Engineering Landmarks, 13, 20, 22, 28, 32, 34, 36, 42, 48, 85, 92, 103–104, 111, 113, 133, 136
Amman, Othmar H., 50
Amtrak. *See* National Railroad Passenger Corp.
Arthur Kill Lift Bridge (N.J.), 62–64
Ashtabula Creek Bridge, 4
Atchison, Topeka & Santa Fe RR, 92, 142–143

Baltimore & Ohio RR, 1, 9–16, 20–22, 62–64, 71–72, 75–76, 118–121, 144–145, 147
Baltimore & Port Deposit RR, 15
Baltimore & Susquehanna RR, 9
Baltimore Bridge Co., 29, 31
Barlow, Peter W., 106–107
Bartlett, W.H., drawing by, 14
Beach, Alfred Ely, 107, 117
Beahan, Willard, 69
Ben Franklin Bridge (Pa.), 52
Bessemer, Henry, 5
Bessemer & Lake Erie RR, 147–150
Birnie, Alexander, 77
Black Rock Tunnel (Pa.) 104
Blue Ridge RR, 104, 109–111
Blue Ridge Tunnel (Va.), 104, 109–111
Bogart, John, 70
Bollman, Wendell, 2, 20–22; bio., 21
Bonzano, Adolphus, 32
Boston & Albany RR, 75–78
Boston & Maine RR, 2, 111–114
Boston & Providence RR, 15–17
Boston & Worcester RR, 75
Bouscaren, Louis G. F., 9, 29; bio., 31
Brayton, Benjamin, 179
bridges, 1–67, 170–185. *See also* names of individual bridges
    failures, 4, 11, 51, 52–53
    foundations: cofferdam, 5, 35–36, 41, 46, 58, 63; open dredging, 38, 60–61; piling, 5; pneumatic caisson, 5, 26–27, 46, 49, 51–52, 54–55; "sand island," 60–61
    materials: cast iron, 2, 7; concrete, 8–9, 45–48; masonry, 1, 9, 12–20, 34–36, 39–42; steel, 5, 26, 37, 51, 65; timber, 1–2, 10, 42–45, 93–94, 170–174; wrought iron, 2, 22–24, 30
    specifications, 5, 11, 29
    types:
        arch, 1, 7–9, 12–20, 26–28, 34–36, 39–42, 45–51
        box girder, 63–67
        cantilever, 7, 28–32, 36–38, 51–54, 59–62
        lift, 62–64
        plate girder, 9
        pontoon, 180–182
        suspension, 6–7, 28–29, 31, 175–177
        trestle, 42–45
        truss, 1–4; Baltimore, 3, 11; Bollman, 2, 20–22; continuous, 8, 57–59; Fink, 2–3, 11; Howe, 2–3, 93–94, 172–173; Pennsylvania, 3, 11; Petit, 3, 11; Pratt, 2, 54–55, 62; simple span, 5–6, 54–56; Town, 1–2; Warren, 2, 32, 57, 60; Whipple, 2, 4, 10, 21, 30; tubular, 22–25
British Columbia Rail, 166, 168
Brooklyn Bridge (N.Y.), 175
Brown, William H., 40
Brunel, Marc Isambard, 106–107
Budd, Ralph, 97
Burleigh, Charles, 112
Burlington Northern, 64–67
Burlington Northern & Santa Fe RR, 86, 92, 99, 168
Burr, William H., 5

caisson disease, 5, 27, 107
Campbell, Henry R., 180
Canadian Government Ry., 51–53
Canadian National Ry., 25, 51–53, 108, 114–118, 166, 168
Canadian Pacific RR, 8, 31, 48, 93–96, 106–109, 121–124, 136–139, 146, 168, 173
Canadian Society for Civil Engineering, 53
Canton Viaduct (Mass.), 15–17
Carnegie, Andrew, 27, 147
Carnegie Steel Co., 147
Carroll, James, 12
Carroll, John, 13
Carrollton Viaduct (Md.), 12–13
Cartlidge, Charles H., 54–55
Cascade Bridge (N.Y.), 170–172
Central Pacific RR, 81–85
Central RR of New Jersey, 8
Chesapeake & Ohio RR, 57–59, 109–111
Chester & Holyhead Ry., 22, 25
Chicago, Burlington & Quincy RR, 21, 39, 54–56, 64
Chicago, Milwaukee, St. Paul & Pacific RR, xii, 74, 96, 142, 180–182
Chicago, Rock Island & Pacific RR, 26, 31, 97, 143, 178–179
Chicago & Alton RR, 5
Chicago & North Western RR, 36, 158–160
Chicago & Rock Island RR. *See* Chicago, Rock Island & Pacific RR
Chihuahua-Pacific Ry., 99–101
Cincinnati, New Orleans & Texas Pacific Ry., 31
Cincinnati Southern RR, 28–32
Clarke, Reeves & Co., 32
Clarke, Thomas Curtis, 4–5, 32, 37, 70; bio., 39

classification yards, 141–144, 151–154, 158–160, 163–166
    Argentine Yard, 143
    Armourdale Yard, 143
    Bailey Yard (Neb.), 143
    Barstow Yard (Calif.), 142
    Conway Yard (Pa.), 154
    Enola Yard (Pa.), 151–154
    Gibson Yard (Ind.), 142
    Honey Pot Yard (Pa.), 142
    Knoxville Yard (Tenn.), 163
    Pigs Eye Yard (Minn.), 142
    Proviso Yard (Ill.), 158–160
    Silvis Yard, 143
    Spencer Yard (N.C.), 163–166
Cleveland, Cincinnati, Chicago & St. Louis RR (Big Four), 5
Cleveland, Pres. Grover, 31
coal docks, 144, 146–147, 161–163
    B&O, Curtis Bay, Md., 147
    B&O RR and NYC RR, Toledo, Ohio, 144
    Lamberts Point, Norfolk, Va., 161–163
    Port Richmond Terminal (Pa.), 146–147
    Sewalls Point, Norfolk, Va., 147
Cohen, Abraham B., 45
Columbia & Kootenay RR, 3
Connecticut River Bridge (Mass.), 10
Consolidated Rail Corp. (Conrail), 20, 41–42, 77–78, 104, 129, 154
Coolidge, Pres. Calvin, 132, 135
Cooper's Loading, 9, 11, 51, 54, 58, 60, 65
Cooper, Theodore, 3, 9, 51; bio., 11
Corbin, Austin, 127
Creel, Enrique, 100
Crocker, Alvah, 112–113
Crocker, Charles, 81–82, 85
Crozet, Claudius, 109–111; bio., 111
CSX Transportation, 13, 15, 59, 121
Cumberland, Stuart, 121
cut and fill, 73

Dale Creek Bridge (Wyo.), 172
Dakota & Great Southern RR, 11
Deadwood Central RR, 70
Delair Bridge (Pa.), 6
Delaware, Lackawanna & Western RR, 9, 45–48
Denver & Rio Grande Western RR, 71–72, 129–133
Denver, Northwestern & Pacific RR, 129
Denver & Salt Lake RR, 130–133
Doane, Thomas A., 112–113; bio., 112
Downing, Robert W., 66
Droege, John A., 141
Duluth, Missabe & Iron Range RR, 155–157
Duluth, Missabe & Northern RR, 145, 154–156

Eads, James Buchanan, 5, 7, 11, 26–28, 31, 107; bio., 28
Eads Bridge. See St. Louis Bridge
Edwards, A. F., 112
Ellet, Charles, Jr., 175
Erie RR, 11, 18–20, 32–34, 170–174

Fink, Albert, 2, 3; bio., 11
Fitchburg RR, 112
Flagler, Henry M., 182–185
Florida East Coast Ry., 182–185
Florida Keys Bridges (Fla.), 182–185
Fox, John, 137

George Washington Bridge (N.Y.), 50
Georgia RR, 79
Gerber, Heinrich, 7

Gilman, L. C., 136
grain elevators, 143
    Great Northern (Wis.), 144–145
    PRR (Md.), 143–145
Grand Trunk RR, 22–25, 107, 114–118, 175–177
Greathead, James Henry, 106–107
Great Northern Ry., 34–36, 64, 71, 74–75, 96–99, 106, 133–136
Great Western Ry., 114
Green River Bridge (Ky.), 3, 11
Greiner, J. E., 8
Guilford Transportation Industries, 113
Gunnison, Capt. John W., 70

Harriman, Edward H., 44
Haskell, C. F. B., 99
Haskin, DeWitt Clinton, 107–108
Haupt, Herman, 3, 80, 112; bio., 10–11
Hawkesbury River Bridge (Australia), 39
Hayes, Pres. Rutherford B., 31
Hell Gate Bridge (N.Y.), 7, 48–51, 57
Heineman, Ben W., 160
High Bridge (Ky.), 7, 28–32
Hill, James J., 35–36, 54, 93–94, 96–97, 133, 136
Hobson, Joseph, 114–118; bio., 114
Hocking Valley RR, 57
Hodges, James, 22–23
Hometown Viaduct (N.J.), 8
Hood, William, 42, 89–90; bio., 92
Hopkins, Mark, 81–82
Horseshoe Curve, Pennsylvania RR. See Mountain passes, Allegheny Mountains
Howard Needles Tammen & Bergendorff, 65
Howe, William, 2; bio., 10
Hudson & Manhattan RR, 107–108
Hudson River Bridge, N.Y.: proposed, 31, 49, 57; at Poughkeepsie, N.Y., 36–38
Huey P. Long Bridge (La.), 59–62
Hulett, George H., 145, 148
Hulett Automatic Unloading Machine, 140, 145–150
hump yards. See classification yards
Hunt, Gov. Washington, 174
Huntington, Collis P., 81–82

Illinois Central RR, 56
Indiana Harbor Belt RR, 142
intermodal terminals, 146–147, 160, 166–169
    Deltaport (B.C.), 166–169
    Wolfe's Cove (P.Q.), 146

Jacobs, Charles Mattathias, 124–127; bio., 127
James River Bridge (Va.), 2
Judah, Theodore Dehone, 70, 81–82, 85; bio., 82

Kansas City, Mexico & Orient Ry., 100
Kelley Creek Bridge (Idaho), xii
Kellogg, Charles, 39
Kellogg, Clarke & Co., 39
Kinnear, Wilson S., 127
Kinzua Viaduct (Pa.), 32–34
Kirkwood, James Pugh, 18–19; bio., 19
Knight, Jonathan, 15–16, 76
Kootenay River Bridge (B.C.), 3
Knox & Kane RR, 34
Krome, William J., 183

Lake Shore & Michigan Southern RR, 4
Land Grants Act of 1862, 71
Latah Creek Bridge (Wash.), 64–67
Latrobe, Benjamin H., 13

Latrobe, Benjamin H., Jr., 11, 13, 21, 31, 75; bio., 15
Lawler, John D., 180–182
Letchworth, W. P., 174
Lewis, George, 132
Lewis and Clark Expedition, 97
Lexington & Danville RR, 28
Lincoln, Pres. Abraham, 179
Lindenthal, Gustav, 6–8, 32, 48–49, 57–58; bio., 50
Linville, Jacob H., 3–4
Little Patuxent River Bridge: at Laurel, Md., 20; at Savage, Md., 21–22
Lloyd, James, 12
Loder, Benjamin, 174
Long, Col. Steven H., 12
Long, Gov. Huey P., 62
Long Island RR, 127
loops or spirals, 74–75; El Lazo (Mexico), 100; Georgetown (Colo.), 74–75; Tehachapi (Calif.), 75, 89–92
Los Angeles County Metropolitan Transit Authority, 108
Louisa RR, 109
Louisville & Nashville RR, 3, 11, 31
Lucin Cut-Off (Utah), 42–45, 92

McCartney, John, 12
McNeil, Anna, 76
McNeil, Capt. William Gibbs, 15–16, 76; bio., 16
Madison RR, 72
Main River Bridge (Germany), 7
Manhattan Bridge (N.Y.), 50
Manhattan Bridge Co., 37
Marent Gulch Bridge (Mont.), 172
Maryland Commuter Rail Service (MARC), 15
Mateos, Pres. Adolfo Lopez, 101
Meredith, Joseph C., 183
Mesabi Range, Minn., 155
Metlac Viaduct (Mexico), 87–89
MetroLink light rail system (St. Louis, Mo.), 27–28
Mexican National Rys. (FNM), 88–89, 101
Mexican Ry. (FCM), 86–89; new, 101
Michigan Central RR, 7, 127–129
Milwaukee & Mississippi RR, 180
Milwaukee & St. Paul RR, 180–182
Mine Hill & Schuylkill Haven RR, 31
Minnesota & Northwestern RR, 128
Minneapolis Union Ry., 35
Mississippi River Bridge: at Quincy, Ill., 21, 39; at Wabasha, Minn., 181–182; at Rock Island, Ill., 26, 31, 178–179; at Memphis, Tenn., 7; at Prairie du Chien, Wis., 180–182; at St. Paul, Minn., 31
Missouri River Bridge: at Glasgow, Mo., 5; at Chamberlain, S.D., 181–182; at Bismarck, Dakota Territory, 4; at St. Charles, Mo., 31
Moberly, Walter, 70, 94
Modjeski, Ralph, 51, 54–55, 59; bio., 52
Moffat, David H., 129–130, 133
Moffat Tunnel Bill, 130
Monocacy River Bridge (Md.), 1
Montgague, Samuel S., 85
Monzani, Willoughby, 2
Morison, George S., 4, 7, 52
Mountain Creek Bridge (B.C.), 93–94, 173
mountain passes
    Allegheny (Pa.), 10, 71, 74, 78–81
    Berkshire Hills (Mass.), 75–78
    Donner (Calif.), 70, 81–86, 92
    Homestake (Montana), 74
    Humboldt (Nev.), 82
    Kicking Horse (B.C.), 94,
    Maltrata Incline (Mexico), 86–89
    Marias (Mont.), 70–71, 96–99
    Mullan (Mont.), 74
    Pipestone (Mont.), 74
    Rogers (B.C.), 93–96
    Rollins (Colo.), 129–130
    Saluda (N.C.), 72
    Sierra Madre Occidental (Mexico), 99–101
    Soldier Summit (Utah), 72
    Stampede Pass (Wash.), 75
    Stevens (Wash.), 70, 75
    Tehachapi (Calif.), 68, 89–92
    Tennessee (Colo.), 71, 129
    Yellowhead (B.C.), 99
Municipal Bridge (St. Louis, Mo.), 5

Nashville, Chattanooga & St. Louis RR, 54, 56
National Historic Civil Engineering Landmarks. *See* American Society of Civil Engineers
National Railroad Passenger Corp. (Amtrak), 8, 16–17, 49, 127; Northeast Corridor, 16–17
National Railways of Mexico, 88–89, 101
National Register of Historic Places, 13, 15, 16, 20, 22, 28, 34, 36, 39, 42, 45, 48, 78, 81, 103–104, 113, 121
New York, Lake Erie & Western RR. *See* Erie RR
New York, New Haven & Hartford RR, 16, 38–39, 48–51
New York & Erie RR. *See* Erie RR.
New York & Susquehanna RR, 20
New York Central RR, 71, 75–78, 126–129, 144
New York Connecting RR, 48–51
Niagara Bridge Co., 175
Niagara River Bridge (N.Y.): Grand Trunk RR, 114, 175–177; Michigan Central RR, 7
Noble, Alfred, 7, 52
Norfolk & Western RR, 161–163
Norfolk Southern Ry., 31, 72, 163, 166
Northern Pacific RR, 4, 11, 64, 74–75, 96, 172
Norwich & Worcester RR, 104

Occoquan Creek Bridge (Va.), 6
Ohio River Bridge: at Bellaire, Ohio, 21; at Sciotoville, Ohio, 8, 50, 57–59; at Cincinnati, Ohio, 4, 31; at Louisville, Ky. (Big Four), 5; at Metropolis, Ill., 5, 9, 54–56; at Louisville, Ky. (PRR), 3, 11
ore docks, 140, 144–150, 154–157
    B&O RR and NYC RR (Toledo, Ohio), 144
    B&O RR (Lorain, Ohio), 145
    Bessemer & Lake Erie RR (Conneaut, Ohio), 147–150
    Duluth, Missabe & Iron Range RR (Duluth, Minn.), 155–157
    Duluth, Missabe & Northern RR (Duluth, Minn.), 145, 154–156
    PRR (Cleveland, Ohio), 140
Osborne, Richard B., 2
Owen, Albert K., 99

Pacific Railroad surveys, 70, 97
Paducah & Illinois RR, 54–56
Panama Canal, 97
Parsons Brinckerhoff Quade & Douglas, 137
Patapsco Bridge & Iron Works, 21
Patterson Viaduct (Md.), 13
Peninsular & Occidental Steamship Co., 183
Pensil Viaduct-Tunnel (Mexico), 89
Penn Central Co., 17, 41, 154
Pennell, Joseph, etchings by, 46, 50
Peto, Brassey & Betts, 22
Pennsylvania RR, 3, 6, 9–11, 39–42, 48, 71–72, 74, 78–81, 103, 107, 118, 124–127, 140, 142–143, 151–154
Philadelphia & Reading RR, 2, 104, 127, 146–147
Philadelphia & Reading RR Bridge (Pa.), 2

Phoenix Bridge Co., 6, 8, 51g
Port Authority of New York & New Jersey, 64
Portage Viaduct (N.Y.), 173–174
Porter, H. K. Co., 72
Portland & Ogdensburg RR, 15
Potomac River Bridge: Va., 1, 20–22; Md., 21
Poughkeepsie-Highland Railroad Bridge Co., 39
Pratt, Caleb, 2
Pratt, Thomas W., 2
Providence & Worcester RR, 104
Public Belt RR Commission, New Orleans, La., 59

Quebec Bridge (P.Q.), 7, 51–53
Quebec Bridge Co., 51
Queensboro Bridge (N.Y.), 50, 57.

railroad curves, 72–73
railroad grades, 71–75; compensated, 72; ruling, 72
railroad location, 69–75
Rappahannock River Bridge (Va.), 8
Ray, George J., 45
Reading RR. *See* Philadelphia & Reading RR.
Reeves, Samuel, 5
Rensselaer & Saratoga RR, 10
retarders, hump yard, 142–143, 152–154, 158, 163–165
Richmond, Fredericksburg & Potomac RR, 6, 8, 11
Richmond & Petersburg RR, 2
Righter, H. R., 40
Rio Chinipas Bridge (Mexico), 100
Rio Fuerte Bridge (Mexico), 100
Roebling, John Augustus, 7, 28–29, 31, 114, 175–177; bio., 175
Roebling, Washington Augustus, 175
Rogers, Maj. Albert B., 93–94, 121, 136; bio., 96
Rogers, David H., 66
Roosevelt, Pres. Theodore, 97
Ross, Alexander McKenzie, 22; bio., 25
Ross, James, 94–95

Sacramento Valley RR, 82
St. Clair Tunnel Co., 116
St. Croix River Bridge (Minn.), 7
St. Gotthard Ry. (Switzerland), 122
St. Lawrence Bridge Co., 51
St. Lawrence River Bridge (P.Q.), 8, 31
St. Lawrence Seaway, 24–25
St. Louis Bridge (Mo.), 5, 7, 11, 26–28, 107
San Francisco–Oakland Bay Bridge, 52
Schlatter, Charles L., 78
Schneider, Charles C., 4, 7
Schwitzer, J. E., 122–124
Secrettown Trestle (Calif.), 83
Seymolur, Silas, 173–174
Shanly, W. & F. & Co., 112–113
Shedd, Jack P., 65
Smith, Col. Charles C. 35
Smith, Charles Shaler, 8, 29–31; bio., 31
Smith, Latrobe & Co., 31
Smithsonian Institution, National Museum of American History, 2
snowsheds, 83, 85, 94–96
Southern Pacific RR, 42–45, 59–62, 68, 81–85, 89–92, 70
Southern Ry., 30–31, 72, 163–166
Spettel, Michael, 180–182

Spokane, Portland & Seattle RR, 64
Stanford, Leland, 81–82
Stanton, Robert, 74
Starrucca Viaduct (Pa)., 18–20, 171
Staten Island Rapid Transit Co. (B&O), 62–64
Stephenson, George, 24
Stephenson, Robert, 22, 24, 114; bio., 25
Stevens, Maj. Isaac J., 97
Stevens, John Frank, 70–71, 97–99, 133, 136; bio., 97
Stilwell, Arthur, 100
Stone, Amasa, Jr., 10
Stone Arch Bridge: Minn., 34–36; Pa., 39–42
Stoney Creek Bridge (B.C.), 94, 139, 173
Strong, Daniel W. "Doc," 81
switchbacks, 74–75

Talcott, Capt. Andrew, 86, 88
Terzaghi, Dr. Karl, 60
Texas & Pacific RR, 62
Thomas, Philip E., 14
Thomas Viaduct (Md.), 13–15
Thomson, John Edgar, 78–81; bio., 79
Town, Ithiel, 1
Troy & Greenfield RR, 11, 111–113
Tunkhannock Viaduct (Pa.), 45–48

tunneling, 102–139
    machines, 108–109, 117–118, 137
    methods, 103–109; American, 106; Austrian, 105–106; shield-driven, 106–108, 115–117, 124–127; soft ground, 118–121; "trench and tube," 107, 127–129
tunnels, 103–139, passim. *See also* names of specific tunnel
    spiral, 75, 121–124
Turner, Claude A. P., 7

Union Bridge Co., 37, 39
Union Pacific RR, 56, 62, 73–75, 85–86, 92, 133, 143, 160, 168, 172
U.S. Supreme Court, 178–179

Valley RR of Virginia, 22
Van Horne, William, 94, 96
VIA Rail, Canada, 24, 52
Victoria Bridge (P.Q.), 22–25
Virginia & Kentucky RR, 111
Virginia Central RR, 104, 109–111
Virginian Ry., 147

War Department, U.S., 49, 54, 57, 59
Warren, James, 2
Warren, Mass., Bridge, 10
Webb, Walter L., 70
Wernwag, Lewis, 1
Western Pacific RR, 73
Western RR of Mass., 10, 16, 19, 75–78, 111
Wever, Caspar W., 12
Whipple, Squire, 2, 3; bio., 10
Whistler, George Washington, 16, 19, 75–78; bio., 76
Whistler, James Abbott McNeil, 76
Wilgus, William John, 127–129; bio., 128
Wimer Viaduct (Mexico), 87–88
Wisconsin Central RR, 7